GRAPHENE OXIDE - APPLICATIONS AND OPPORTUNITIES

Edited by **Ganesh Shamrao Kamble**

Graphene Oxide - Applications and Opportunities
http://dx.doi.org/10.5772/intechopen.73080
Edited by Ganesh Shamrao Kamble

Contributors

Efim Neustroev, Biju Majumdar, Daisy Sarma, Tridib Kumar Sarma, Maksim Gudkov, Valery Melnikov, Guilherme Cordeiro, Lingwen Zeng, Shi-Lin Cao, Hang Yin, Jun Xiong, Dong-Hai Lin, Jorge Uruchurtu, Carmina Menchaca, Cesar Garcia Perez, Elsa Pereyra Laguna, Miriam Flores, Miguel Angel Garcia Sanchez, Ali Hassan Gemeay, Mohamed E. El-Halwagy, Joong Tark Han, Ganesh Shamrao Shamrao Kamble

First published in London, United Kingdom, 2018 by IntechOpen
IntechOpen is the global imprint of INTECHOPEN LIMITED, registered in England and Wales, registration number: 11086078, The Shard, 25th floor, 32 London Bridge Street
London, SE19SG – United Kingdom
Printed in Croatia

British Library Cataloguing-in-Publication Data
A catalogue record for this book is available from the British Library

Additional hard copies can be obtained from orders@intechopen.com

Graphene Oxide - Applications and Opportunities , Edited by Ganesh Shamrao Kamble
p. cm.
Print ISBN 978-1-78923-588-3
Online ISBN 978-1-78923-589-0

Contents

Meet the editor

Dr. Ganesh S. Kamble completed his PhD degree in 2011 in Chemistry from the Department of Chemistry, Shivaji University, Kolhapur, India. Most of his research focuses on analytical and material chemistry. After completing his PhD degree, he worked as an assistant professor at the Department of Engineering Chemistry, Kolhapur Institute of Technology's College of Engineering (Autonomous), Kolhapur, Maharashtra, India. He is a life member of the Association of Separation Scientist and Technologists [ASSET], BARC, Mumbai, India. He is a reviewer of peer-reviewed international journals of many publications. He received a "Summer Research Fellow-2013" award from the Indian Academy of Sciences (IASc), Bangalore. He also received a "Postdoctoral Research Fellowship" from the Ministry of Science and Technology (MOST), P.R. China. He is currently working as a postdoctoral research fellow at the Department of Chemistry, National Tsing Hua University, Taiwan, since November 2017.

For EU product safety concerns:
IN TECH d.o.o., Prolaz Marije Krucifikse Kozulić 3, 51000 Rijeka, Croatia,
info@intechopen.com or visit our website at intechopen.com.

Preface

Nowadays, the world is facing challenges in the field of life science, energy, health, and the environment with enactment, healthiness, and toughness materials. It has been widely distinguished that advanced materials including carbon-based nanomaterials such as carbon nanotubes (CNTs), carbon blocks, and graphene oxide (GO) will continue to play a significant role in overcoming the major challenges and making the innovations in both technology and practical application. Owing to the prolonged structural properties, GO and its nanocomposites hold unlimited promises that have been much more widely attracted in various important and versatile applications such as catalysis, energy conversion and storage devices, hydrogen storage materials, various lithium ion batteries, sodium ion batteries, zinc air batteries, sensing electronic devices, environment protection, photocatalyst for water splitting, water purification, and removal of air pollutants. The outstanding potential of graphene-based nanomaterials has enabled significant enrichments for abundant imperative applications.

This book that gives a comprehensive information is envisioned to address the sustained progresses and challenges with a wide scope of attention, highlighting the fundamental understanding of the synthesis and characterization and the performance of graphene and functionalized graphene oxide composite materials. To assist further technical challenges, practical applications and prospects of graphene and its composite materials are discussed, and numerous future research directions are also suggested in this book.

Dr. Ganesh S. Kamble
Department of Engineering Chemistry
Kolhapur Institute of Technology's College of Engineering (Autonomous)
Kolhapur, India

Department of Chemistry
National Tsing Hua University
Hsinchu, Taiwan

Section 1

Graphene Oxide Synthesis and Applications

Chapter 1

Introductory Chapter: Graphene Oxide: Applications and Opportunities

Ganesh Shamrao Kamble

Additional information is available at the end of the chapter

http://dx.doi.org/10.5772/intechopen.79640

1. Introduction

This chapter aims to introduce the emerging technologies of graphene oxide (GO) in various fields such as industrial, medical, electronics, artificial intelligences, materials and alloys, energy storage devices, optical, physics, mechanical, nanomaterials, and sustainable chemistry. Graphene oxide analogy to graphene was first discovered by chemist Benjamin C. Brodie in 1859 and further quick method was developed by Hummers and Offeman in 1957; globally, the method is known as Hummers' method [1].

2. History of synthesis of GO and structure

Graphene is a two-dimensional (2D) carbon sheet having sp^2 hybridization with molecular weights of more than 106–107 g/mol. It has been packed into a honeycomb lattice (**Figure 1**). The bulk material of graphite that was discrete in single monolayer sheets showed noteworthy properties and hence its single monolayer structure motivated in various applications. The exfoliation of graphene oxide was synthesized by using strong oxidizing agents such as $KMnO_4$ and conc. H_2SO_4 [2, 3].

3. Overview of applications and future opportunities of GO

Many devices of GO overtake reference systems, for example, capacitors [4, 5], foldable electronic devices [6], translucent electrodes [7], biomedical applications [8], pollution management [9], sensors [10], H_2-generation [9] and energy applications [11].

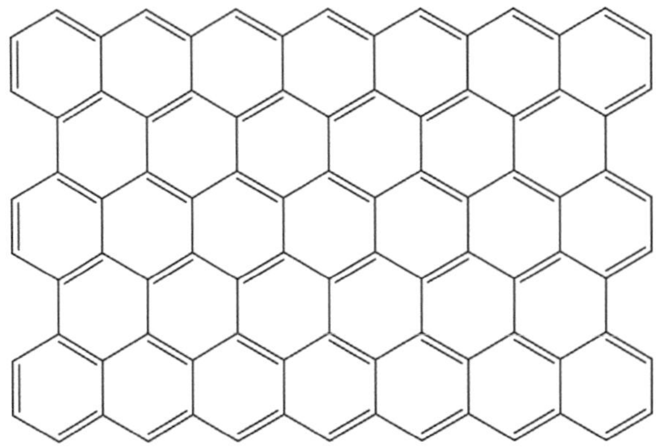

Figure 1. Schematic representation of single layer graphene oxide with zig-zag and arm-chair edges.

Because of its honeycomb lattice with two carbon atoms per unit cell, graphene oxide shows an innumerable of exceptional chemical and physical properties. Due to the valence band and conduction band touch, the Brillouin zone corners [12] so as charge carriers in graphene behave like massless relativistic particles. Due to the delocalized out-of-plane π bonds arising from the sp^2 hybridization carbon atoms, an unprecedented high carrier mobility of $\approx$200,000 $cm^2\ V^{-1}\ s^{-1}$ has been achieved for suspended graphene [13].

For the bulk production of GO, exfoliation is the most developed attractive method. The pristine graphite is converted into graphite oxide (GO sheets) by using a mixture of $KMnO_4$ and concentrated H_2SO_4 [14–16]. In the oxidation of GO, large numbers of oxygen-containing functional groups such as epoxides, carboxyl and hydroxyl groups are attached onto the graphene basal plane and edges. Due to its hydrophilic nature, it is easily dispersed in water or polar organic solvents. The structural and electrical properties of pristine graphene are obtained by using reducing agents and thermal treatment, sodium borohydride [17], hydrazine [18] and thermal reduction [19, 20], respectively. Due to carcinogenic and highly toxic reducing agents property, in the recent years, reduction of GO is carried out by green reductants agents such as polyphenols of green tea, melatonin, vitamin C, bovine serum, albumin, sugars and even bacteria was also studied. Hydrothermal, solvothermal reduction, catalytic and photocatalytic reductions have also been developed. Furthermore, surfactant and boiling point of solvents also effect on GO.

At the current level of development, the properties and binding structure of graphene are important toward the recent applications. The knowledge produced by the systematic functionalization of graphene could be a much haunting basis for discovering the chemistry and nanomaterials.

Finally, GO and GO-based nanomaterials and its graphene derivatives are essential for future applications such as fuel cells, vivo sensors, supercapacitors, energy storage devices, and transparent electronics, which will undoubtedly improve when defined graphene derivatives are employed. Future technology expected that the full development and growth will depend only on graphene and its functionalized composite materials. This chapter highlights the challenges and opportunities associated with GOs. Subject of interest in this chapter is exploring opportunities and technologies related to energy, pure water and good health.

Author details

Ganesh Shamrao Kamble[1,2]*

*Address all correspondence to: ganeshchemistry2010@gmail.com

1 Kolhapur Institute of Technology's, College of Engineering (Autonomous), Kolhapur, India

2 Department of Chemistry, National Tsing Hua University, Taiwan, Hsinchu, Taiwan

References

[1] https://en.wikipedia.org/wiki/Graphite_oxide#History_and_preparation

[2] Croft RC. Lamellar compounds of graphite. Quarterly Reviews, Chemical Society. 1960; **14**:1-45

[3] Hummers WS, Offeman RE. Preparation of graphite oxide. Journal of the American Chemical Society. 1958;**80**:1339-1339

[4] Huang Y, Liang J, Chen Y. An overview of the applications of graphene-based materials in supercapacitors. Small. 2012;**8**:1805-1834

[5] Li J, Östling M. Prevention of graphene restacking for performance boost of supercapacitors. Crystals. 2013;**3**:163-190

[6] Chen H, Guo X. Field-effect transistors: Unique role of self-assembled monolayers in carbon nanomaterial-based field-effect transistors. Small. 2013;**9**:1144-1159

[7] Eigler S. A new parameter based on graphene for characterizing transparent, conductive materials. Carbon. 2009;**47**:2936-2939

[8] Chung C, Kim YK, Shin D, Ryoo SR, Hong BH, Min DH. Biomedical applications of graphene and graphene oxide. Accounts of Chemical Research. 2013;**46**:2211-2024

[9] Xie G, Zhang K, Guo B, Liu Q, Fang L, Gong JR. Graphene-based materials for hydrogen generation from light-driven water splitting. Advanced Materials. 2013;**25**:3820-3839

[10] Schedin F, Geim AK, Morozov SV, Hill EW, Blake P, Katsnelson MI, Novoselov KS. Detection of individual gas molecules adsorbed on graphene. Nature Materials. 2007;**6**:652-655

[11] Lü K, Zhao G, Wang X. A brief review of graphene-based material synthesis and its application in environmental pollution management. Chinese Science Bulletin. 2012;**57**: 1223-1234

[12] Novoselov KS, Geim AK, Morozov SV, Jiang D, Katsnelson MI, Grigorieva IV, Dubonos SV, Firsov AA. Two-dimensional gas of massless Dirac fermions in graphene. Nature. 2005;**438**:197-200

[13] Du X, Skachko I, Barker A, Andrei EY. Approaching ballistic transport in suspended graphene. Nature Nanotechnology. 2008;**3**:491-495

[14] Stankovich S, Dikin DA, Dommett GHB, Kohlhaas KM, Zimney EJ, Stach EA, Piner RD, Nguyen ST, Ruoff RS. Graphene-based composite materials. Nature. 2006;**442**:282-286

[15] Sudibya HG, He Q, Zhang H, Chen P. Electrical detection of metal ions using field-eEffect transistors based on micropatterned reduced graphene oxide films. ACS Nano. 2011;**5**: 1990-1994

[16] Stankovich S, Dikin DA, Piner RD, Kohlhaas KA, Kleinhammes A, Jia Y, Wu Y, Nguyen ST, Ruoff RS. Synthesis of graphene-based nanosheets via chemical reduction of exfoliated graphite oxide. Carbon. 2007;**45**:1558-1565

[17] Shin H-J, Kim KK, Benayad A, Yoon S-M, Park HK, Jung I-S, Jin MH, Jeong H-K, Kim JM, Choi J-Y, Lee YH. Efficient reduction of graphite oxide by sodium borohydride and its effect on electrical conductance. Advanced Functional Materials. 2009;**19**:1987-1992

[18] Kim MC, Hwang GS, Ruoff RS. Epoxide reduction with hydrazine on graphene: a first principles study. The Journal of Chemical Physics. 2009;**131**:06470

[19] Yang D, Velamakanni A, Bozoklu G, Park S, Stoller M, Piner RD, Stankovich S, Jung I, Field DA, Ventrice CA Jr, Ruoff RS. Chemical analysis of graphene oxide films after heat and chemical treatments by X-ray photoelectron and micro-raman spectroscopy. Carbon. 2009;**47**:145-152

[20] Chen W, Yan L, Bangal PR. Preparation of graphene by the rapid and mild thermal reduction of graphene oxide induced by microwaves. Carbon. 2010;**48**:1146-1152

Chapter 2

Plasma Treatment of Graphene Oxide

Efim Petrovich Neustroev

Additional information is available at the end of the chapter

http://dx.doi.org/10.5772/intechopen.77396

Abstract

Plasma treatment of graphene oxide (GO) is of interest for many applications such as gas sensors, flexible electrodes, biological applications, supercapacitors, batteries, etc. Plasma treatment is a high-tech process of processing materials in combination with efficiency, economy and environmental friendliness. At the same time, plasma treatments can lead to an increase in the defectiveness of the surface of GO. Therefore, it is important to know how the treatment in various gas media affects the properties of GO. This is necessary for the correct selection of processing parameters in order to reduce the negative impact of plasma treatment. The review presents the results of experimental studies of the effect of plasma in gases of oxygen, nitrogen, ammonia, sulfur hexafluoride, carbon tetrafluoromethane, hydrogen and methane on the properties of GO. Plasma treatments were used for the functionalization, reduction, doping of graphene oxide, and also for the obtaining GO from graphene by oxidation in oxygen plasma. The effects of plasma treatment are largely determined by the type of ions used and processing conditions: plasma power, processing time and temperature, pressure, as well as the location of the samples in the reaction chamber and their distance from the plasma ignition zone.

Keywords: plasma treatment, oxygen plasma, nitrogen plasma, ammonia plasma, sulfur hexafluoride plasma, carbon tetrafluoromethane plasma, plasma effect on graphene oxide properties

1. Introduction

For several decades, plasma technologies are widely used in the fabrication of microelectronic structures. The main directions of the use of plasma treatments are cleaning of the surface of materials, etching, deposition and modification of the surface properties of materials (changes in hydrophilic, adhesion, conductivity and other properties). With the development of nanotechnologies, plasma methods have found application in the growing and functionalization of nanomaterials, providing unique properties in combination with the economy and environmental safety in the production process. Thus, plasma technologies are widely used both for the growth of nanomaterials by the method of plasma enhanced chemical vapor deposition (PECVD) and for the modification of surface properties. A low-temperature radiofrequency (rf) plasma containing chemically active particles provides unique possibilities for this. For the functionalization and reduction of graphene oxide, various thermal, chemical, optical, and other methods are used [1–5]. Plasma treatment is effective, eco-friendly and low cost method of functionalization and reduction of graphene oxide (GO). Plasma treatment of GO is of interest for the creation of various devices such as gas sensors [6, 7], photosensitive devices [8], biological sensors [9–11], flexible electrodes [12, 13] and et al. Various methods are used to perform plasma treatments [14, 15], which can be conditionally divided into actions with "low power" and "remote" [14], also low and atmospheric pressures [15]. This largely determines the effect of plasma on the properties of graphene oxide. The most frequently used for the functionalization of graphene oxide are treatments in plasmas of nitrogen, ammonia, oxygen, methane, hydrogen and fluorine. The treatment in each of these plasmas has features related to the chemical activity of ions, activated by the action of ultraviolet radiation, by electrons and other active spices of plasma. At the same time, it should be noted that most of the work devoted to the effect of plasma on carbon nanomaterials is associated with the study of the properties of graphene and carbon nanotubes. Despite the fact that exposure in plasma leads to an effective modification of the properties of graphene oxide, a limited number of studies have been devoted to the study of its properties. Below we will consider the effects of plasma of various gases on the properties of GO, taking into account the processing conditions.

2. Oxygen plasma

Oxygen plasma treatment of graphene oxide can be used to etch surface layers [9, 15–19], to functionalize [6, 9, 20] or to obtain GO from graphene by oxidation [21–25]. On the other hand, the oxygen plasma is very chemically aggressive, usually leading to significant changes in the structural and electronic properties of graphene: a high degree of disorder is induced in the graphene lattice even at low power and after a very short exposure time [26]. At the same time, the intensity of defect formation under the influence of oxygen plasma can be significantly reduced by using mild plasma treatments, which use small power, remote location from the plasma source of samples, the use of protective filters and grids [9, 21, 22, 26–29]. As measurements by the XPS method have shown, the effect of low-pressure O_2 plasma on reduced graphene oxide (rGO) leads to the introduction of different oxygen-containing groups [20, 30]. Similar results were obtained when graphene was exposed to oxygen plasma [9, 15, 21, 22]. Oxygen plasma treatment introduces epoxy (C-O-C) and carboxyl (C-OH) groups in the basal plane and edges of graphene, with the epoxy group being the most energetically favorable [1, 8, 17]. This process is accompanied by a decrease in the carbon regions of sp^2-hybridized bonds [27, 30]. Detailed high resolution X-ray photoelectron spectroscopy (XPS) rGO studies after oxygen plasma treatment with a power of less than 50 W have shown that a monotonous increase in the amount of oxygen and a decrease in the carbon content with increasing processing time does not occur [31]. The content of C-O (epoxy, hydroxyl) groups was maximum after a plasma exposure for 1 min, further decreased after treatment for 5 and 10 min. In contrary, the atomic content of C(O)O (carboxyl/lactone) groups increased at the same processing times. The authors explain this by the fusion of carboxyl groups formed at short plasma treatment times and the formation of lactone groups. This process is accompanied by the loss of one oxygen atom, which leads to a decrease in the oxygen content after 5 min of treatment. With the formation of lactone groups, the authors associate the appearance of new absorption lines in IR spectra at wavelengths at 1730 and 1436 cm^{-1}. It should be noted that in the IR spectra treated in GO plasma, a broad band at 3440 cm^{-1} bound to hydroxyl groups is significantly enhanced [20, 32].

The main peaks of Raman spectra of graphene oxide are G band at ~1580 cm^{-1} wavelength and D band at 1530 cm^{-1} [18, 33]. The band G is the result of intraplane oscillations of the bound atoms of sp^2 whereas the D band is due to non-planar oscillations due to the presence of structural defects. In the presence of defects, vacancies and disorderings in the sp^2-domain of the structure graphene, are significantly increases the intensity of the D peak [18, 33]. When graphene is exposed in plasma, the intensity of the D band (I_D) Raman spectra are significantly increased [26, 27, 34, 35]. Even a short-term O_2 plasma exposure for up to 5 s at a power of 2 W leads to a sharp increase I_D [34]. The change in Raman intensities associated with the increase in defects occurs also for rGO treated in oxygen plasma [9, 20, 32]. These changes are associated with an increase in the content of oxygen groups and defects formed during plasma oxidation. An increase of the D band intensity is also associated with the formation of sp^3-hybridized bonds [9, 22].

An investigation of the electrical conductivity of rGO treated in O_2 plasma showed its decrease both with increasing plasma power and processing time [30]. At high powers and long processing times it becomes impossible to measure the conductivity because it was below the measurement range of the instrument (<0.001 $Sxcm^{-1}$). Scanning electron microscopy (SEM) measurements showed that most of the rGO disappears [30]. The authors explain this by the fact that the oxygen plasma destroys sp^2 domains in rGO and introduces various oxygen groups to the sites of rGO defects, which leads to etching of rGO with conversion him to amorphous carbon. At the same time, under 'gentle' processing conditions, it is possible preserve the electrical conductivity of graphene oxidized by plasma [22]. It should be noted that the production of GO from graphene by the plasma treatment method makes it possible to obtain GO with an almost undamaged surface, high hydrophilicity and increased adhesion of the surface [21–23].

Gokus et al. show that strong photoluminescence (PL) can be induced in single-layer graphene using an oxygen plasma treatment [36]. The samples are then exposed to oxygen/argon (1:2) rf plasma (0.04 mbar, 10 W) for increasing time (1–6 s).

Surface morphology studies after low-pressure oxygen plasma at 50 W showed that the surface of the GO becomes more porous and corrugated [9, 37]. This leads to an increase in its wettability for both graphene oxide [20] and graphene [22, 27], treated in oxygen plasma. The increase of the wettability allows to improve its surface reactivity with respect to biomolecular interactions [24]. Oxygen-plasma-treated rGO surfaces were employed as reactive interfaces for the detection of amyloid-beta (Aβ) peptides, the pathological hallmarks of Alzheimer's disease, as the target analytes [9]. Zhao et al. in [6] proposed a chip based gas sensor NH_3 with oxygen plasma treated GO surface. Owing to the large surface-to-volume ratio of GO and the rich chemical groups on its surface and edges, the sensitivity of the sensor to gas molecule absorption was improved. The response was further improved by oxygen plasma treatment on GO film by introducing numerous site binding defects.

3. Nitrogen and ammonia plasma

In many works, gases NH_3 [38–40] and N_2 [32, 41–43] are used to treat GO in the nitrogen containing plasma. Kim and et al. for nitridation of rGO used NH_3 inductively coupled plasma with a power of 10 W at a pressure of 100 mTorr [38]. The content of many oxygen groups (such as epoxy, hydroxyl, carbonyl and carboxyl) attached to GO films was significantly reduced after treatment with NH_3 plasma, and instead of them, C-N bonds were introduced. The ratios of the nitrogen to carbon concentrations (N/C) and oxygen to carbon concentrations (O/C) in less than 10 min became approximately 10 and 23% and were gradually saturated [38].

Nitrogen forms pyridine (pyridinic-N), pyrrole (pyrrolic-N), and graphite configurations (quaternary, graphitic-N) with carbon atoms in the graphene lattice [36, 44–47]. Pyridinic-N is bonded to two carbon atoms of the hexagonal graphene cell on the edge of vacancy-type defects and introduces 1 p-electron into the π-system; pyrrolic-N introduces 2 p-electrons into the π-system and is connected to two graphene atoms of the pentagonal cell; graphitic-N replaces the carbon atom in the hexagonal ring of graphene [36, 44, 47]. The pyridine and pyrrole configurations of nitrogen form a p-type conductivity, while graphitic N ones increase the electron concentration [44, 48]. Kim et al. [38] showed that when doping in graphene NH_3 plasma, pyridine and pyrrole configurations of nitrogen are predominantly introduced, and the ratios N/C and O/C atoms are inversely related to the processing time in the plasma. The same results were received in Refs. [39, 40]. In the work of other authors [32], the concentration of pyridinic-N (48%) was found to be higher than the Pyrrolic-N (29%) and Graphitic-N (15%). At the same time, treatment with N_2 plasma can lead to an increase in the amount of oxygen [10, 47].

Kim et al. [38] is shown that an increase in the electrical conductivity occurs with an increase in the exposure time in NH_3 plasma with a simultaneous decrease in the optical transmission coefficient. The authors explain the increase in the conductivity from 100 S/m to 1666 S/m rGO films by the combination of the effect of nitrogen doping and the reduction of oxygen. Decreasing of the transmittance is attributed to the restoration of electronic conjugation in rGO film. The surface roughness of the films becomes smoother until the processing time reaches 10 min and further substantially does not change. The main cause of this phenomenon can be explained from the point of view of removing organic impurities in the film, reducing the functional groups of oxygen and extracting the sp^2-carbon domains [38].

In [39] the plasma treatment with ammonia of GO and RGO in the reaction chamber with parallel plasma-enhanced chemical vapor deposition (RECVD) diodes was carried out. The conditions of plasma treatment were as follows: plasma power of 200 W at 13.56 MHz, a gas flow rate of 400 sccm, the substrate temperature 150°C, treatment time 1–5 min. Under these conditions, the level of doping with nitrogen in 6% was reached. Comparison of the intensities of the N1 s peaks associated with the formation of bonds C-N in XPS spectra of GO and RGO after plasma treatments in NH_3 showed that the intensity of this peak in RGO than in GO. The authors explain this by the formation of C-N bonds in the interaction of oxygen groups with ammonia. The intensity ratio of Raman peaks I_D/I_G has a nonmonotonic dependence on the processing time. At initial treatment times (up to 1 min) this ratio increases, then reduces and further gradually enhances with increases exposure time in the plasma. The authors explain the decrease in the I_D/I_G ratio by the formation of an intermediate chemical species, such as hydrazine radicals. The increase can be attributed to the restoration of the sp^2 bonds in the GO sheets due to the NH_3 plasma, which is consistent with the XPS results [39]. Studies of surface morphology, carried out in the same work, are consistent with these results. Studies of surface morphology, carried out in the same work, are consistent with these results. Measurements by atomic force microscopy (AFM) showed that exposure to NH_3 plasma for 1 min leads to smoothing of the GO surface. With an increase in the processing time to 5 min, the inhomogeneity of the surface increases due to the destructive effect of the plasma. Four-probe sheet resistance measurements showed that initial treatment with a duration of 1 min leads to a sharp decrease of resistance by 6 orders of magnitude. The resistance reaches a minimum after 3 min of plasma treatment (67.5 ± 4.5 kΩ/sq) and further is a gradual increase of the resistance is observed. The authors attribute the increase in electrical conductivity to the removal of oxygen functional groups and nitrogen doping of graphene oxide.

Kumar and others investigated the effects of plasma N_2 and H_2 (50 sccm each) at a power of 500 W for a time of 1 hour on the GO properties [41]. The microwave plasma source was remoted from the GO sample and the temperature was raised only by 10°C during plasma treatment. After plasma treatment, a slight increase in the Raman intensity ratio of the I_D/I_G peaks from 0.97 to 1.05 was observed. From the XPS data it was obtained that the C/O ratio increases from 2.2 to 5.2. It was found that the nitrogen introduced during the plasma treatment was 5.8 at.% of the material. The decrease in oxygen group content was confirmed not only from XPS measurements, but also from Fourier-transform infrared spectroscopy (FT-IR) spectroscopy data, as well as from other works, for example see [38]. After exposure in plasma, the intensities of the FT-IR peaks corresponding to the oxygen functionalities, such as the C=O stretching vibration peak at 1726 cm^{-1}, was decreased dramatically.

In [9] GO was processed in an N_2 inductively coupled plasma (ICP) (P = 50 W, 50 mTorr, flow rate of 10 sccm). As a result of plasma exposure, the ratio of I_D/I_G Raman intensities of the samples increased from 0.851, which corresponds to graphene oxide, to 1.08. The authors note that when comparing treatments in nitric and oxygen plasmas, in the latter case, significant surface distortions are observed due to the high-energy particles present in the oxygen plasma. From the results of measurements of the FT-IR spectra, it was found that treatment in plasma N_2 leads to a significant decrease of the absorption band corresponding to the O-H group, in contrast to the O_2 plasma treatment. Also in the IR spectra new peaks appear at 1331 cm^{-1}, which confirms the presence of the amide functional group corresponding to N-H in-plane stretching, and the peaks 1608 cm^{-1} belong to C-N bond stretching.

Lee and et al. used inductively coupled NH_3 plasma with a power of 1×10^3 W/m^2 at a pressure of 500 mTorr [40]. Samples were processed in the two plasma regions: the bulk plasma region (R_{bulk}) and the sheath region (R_{sheath}). In both regions, reduction and nitridation processes began immediately once the NH_3 plasma was exposed to the GO films. Just like in Kim's work XPS measurements showed that in both cases a gradual increase N-pyrrolic and a decrease N-quoternary with an increase of treatment time of up to 30 min were observed. The authors also observed an increase in the ratio of N/C and a decrease of the O/C, which can be explained by the substitution of nitrogen compounds at the sites of oxygen functional groups on the r-GO films. At the same time, the electrical conductivity of the r-GO films in the bulk plasma region increased significantly after 10 min of treatment. On the other hand, the optical transmittance of the r-GO films in the R_{bulk} decreases gradually with increasing processing time, while for the sheath plasma region it first decreases, then after 5 min of processing starts to increase gradually. The observed effects are attributed to the fact that the reactions in each region were shown to be different. The authors explain the observed effects by the difference in the reactions in these regions. In the R_{sheath}, the physical reaction was dominant because of the accelerated ion bombardment by the strong electric field. In general, comparing these sample locations in the reaction chamber, the authors conclude that the reduction in the R_{bulk} is more effective.

Mohai et al. in [42] estimated the penetration of nitrogen and argon ions into the GO using the stopping and range of ions in matter (SRIM-2013) program. Calculations have shown that at energies of 20–50 eV, the depth is equal for the two ions. Thus, plasma can only modify narrow near-surface regions.

It was shown in [43] that, as a result of nitrogen plasma treatment, the defect formation causing an increase in the intensity of the Raman peak D of the spectra depends of location of the samples in the reaction chamber. The substrate was placed "face down" into the plasma chamber (**Figure 1**), which significantly reduced the formation of defects in the rGO. In the investigated rGO samples, the electrical resistance increased, it is possible that this is due to the predominance of defect formation over doping and restoration of the graphene lattice.

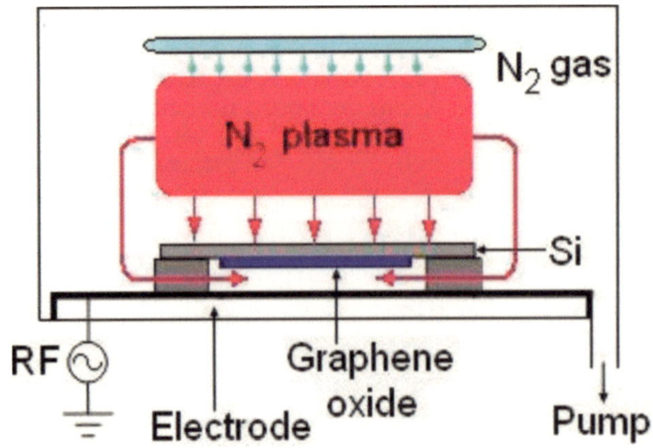

Figure 1. Schematic view of plasma system used for nitridation of graphene oxide.

Qin et al. [49] demonstrated that N_2 plasma treatment is an efficient technique to prepare rGO with simultaneous introduction of N doping and ferromagnetism. The structural characterizations clearly demonstrate that the pyrrolic N bonding configuration is the main source of ferromagnetic moments. At the same time, the N_2 plasma exposure time plays the key role in tuning the magnetization of nitrogen doped rGO.

Recently Wang et al. have shown that the nitrogen plasma treatment is used to modify graphene oxide to enhance the oxygen reduction reaction (ORR) performance, which implicates a burst open and highenergy electron/ion collision mechanism for doping and exfoliating [50].

Zhu et al. [51] devoted much attention to the study of the electrical properties of graphene oxide exposed to ammonium and hydrogen plasmas with the addition of argon in the ratio (9:1). Field effect transistors (FET) of rGO were manufactured for the study of electrical characteristics. It was found that the action of ammonium plasma for 8, 5 min leads to electronic conduction. It was found that the action of ammonium plasma for 8, 5 min. Leads to an n-type conductivity with an electron mobility of 5.43 $cm^2\ V^{-1}\ s^{-1}$. At the same time, with a processing time of up to 5.5 min, the p-type conductivity predominates with a hole mobility of 2.1 $cm^2\ V^{-1}\ s^{-1}$. At intermediate exposure times, ambipolar conductivity is observed. The results of the work performed show that treatment with nitrogen plasma leads to the reduction of graphene oxide due to the removal of oxygen containing groups, as well as to n-doping. From the analysis of the XPS spectrums, the authors believe that nitrogen atoms in the pyridine configuration are responsible for the formation of n-type conductivity. With a short plasma exposure time, oxygen doping dominates, leading to a p-type conductivity of FET transistors.

4. Plasma fluorination

Plasmas of carbon tetrafluoromethane (CF_4) and sulfur hexafluoride (SF_6) are more often used for the fluorination of GO by plasma [8, 52–54]. In [8], the effect of plasma treatment of SF_6 and CF_4 ions on GO in the structure of an organic solar cell was investigated by methods of Raman spectroscopy, XPS, ultraviolet (UV) and infrared (IR) spectroscopy, and photoelectric characteristics measurements. For this purpose, reactive ion etching in a plasma with a power of 20 W with a duration of 10 to 60 s at a pressure of 20 mTorr was used. The results of the research showed that the use of GO films functionalized in the SF_6 and CF_4 plasma makes it possible to increase the conversion efficiency of solar energy from 0.56 to 2.72%. And the best result was achieved by processing in plasma SF_6. From measurements of Raman spectra it follows that the ratio of the Raman spectra of I_D/I_G spectra as a result of plasma fluoride treatment increases slightly (from 1.17 to 1.21). This means an increase in disturbances in the graphene lattice. The XPS method revealed the presence of two peaks of C1s and F1 s in the spectra. C1s peak authors are considered responsible for the formation of C-F and C-F_2 bonds with energies at 288.7 and 290.9 eV after SF_6 and CF_4 plasma treatment. The peak F1 s shows the presence of two components of the C-F bond with energies of 685.4 and 688.1 eV corresponding to the semimetallic and semiconductor bonds. In this case, as the authors believe, fluorine should change the electrostatic potential on the outer surface. The large difference in the electronegativity between carbon (2.55) and fluorine (3.98), according to the authors, means that the C-F bonds on the surface are polar. This is equivalent to introducing a layer of dipoles over the entire surface, which can increase the electrostatic potential energy on the surface with covalent bonds [8].

Zhou et al. used XPS, X-ray diffraction (XRD), AFM and transmission electron microscopy (TEM) methods to study GO obtained by the Hammers method [52]. The CF_4 plasma treatment conditions were as follows: gas flow rate of 1.5 l/h, operating pressure of 20 Pa, a bias voltage of 200 V, the power of 240 W, and process duration of 1, 3, 5, 10, 15, 20 min, respectively. From XPS data it is shown that fluorination in plasma of CF_4 leads to the formation of C-F, C-F_2 and C-F_3-bonds. Moreover, the intensity of the peak (C-F) increased at small times (up to 10 min) of treatment, and then sharply decreased. As the authors believe, this decrease may have resulted from the reduction in the conjugated π-domains. At the same time, the proportion of components CF_2 and CF_3 only increases with increasing processing time. On the other hand, the content of C-C and C-O decreases. The fluorination of GO can occur due to the substitution of hydrogen atoms in C-H and O-H bonds, as well as oxygen atoms in oxygen-containing groups. Using AFM and TEM, a thinning effect for GO exposed to plasma in CF_4 was observed. The thickness of pristine GO after plasma exposes decreased from ~1.9 to 1.3 nm. The hydrophobicity of GO showed that with increasing treatment time to 20 min, super hydrophilic GO becomes neither hydrophilic nor hydrophobic [52].

In Ref. [53] studied the effect of an inductively coupled Ar/SF_6 plasma with a power from 100 to 250 W at a pressure of 15–30 Pa on the property of graphene oxide. Using the X-ray energy dispersive spectroscopy, the fluorine content on the GO surface was found to increase depending on both the plasma power and the fluorination time. In this case, the ratio of the fluorine to oxygen concentrations (F/O) ratio content changes more rapidly than the change in fluorine to carbon concentrations (F/C). This can mean that fluorination is more due to the displacement of oxygen than carbon atoms. There is also a slight change in Raman spectra after plasma treatment lasting up to 30 min. Measurements of the volt-ampere characteristics showed a gradual increase in resistance with increasing processing time.

Yu et al. in [54] the high density plasma etcher was employed to carry out the fluorination at room temperature with the CF_4 flow rate fixed at 20 sccm and the pressure kept at about 0.16 Pa. XPS measurements showed that after fluorination the C/F ratio increased from 17.2 to 27%, depending on the treatment time. The study of surface morphology by AFM showed that the structure of the graphene substrate is well preserved after the plasma treatment, implying negligible damage for the samples. These data are confirmed by the Raman spectra data, from which it follows that the ratio of the peak intensities D (at 1340 cm^{-1}) G (at 1601 cm^{-1}) and the ratio of these two peaks (I_D/I_G) all remain practically the same after fluorination. At the same time, unlike the work of [51], the thickness of the grapheme substrate increases from ~0.4 to ~1.0 nm, which the authors associate with fluorination. Perhaps this difference is due to the short processing times, which in this work were up to 20 s, while in Zhou's work—up to 20 min. Studies of electrical properties after treatments with a duration of up to 10 s showed that the current–voltage (IV) characteristics remained practically unchanged, which indicates, in the opinion of the authors, that the structure of rGO was not damaged by plasma treatment. After the fluoridation of GO authors noted a weak p-doping accompanied by a decrease in mobility due to the appearance of new scattering sites associated with F-containing groups.

The study of the fluorination of various carbon materials by plasma makes it possible to understand more fully the effect of such treatment on the properties of graphene oxide. The effect of plasma SF_6 fluorination on the properties of graphene was investigated in [55–59]. Baraket et al. the chemical vapor deposition (CVD) graphene was treated with a pulsed Ar/SF_6 plasma (pulse duration 2 ms and period 20 ms, plasma exposure time was 6 s and the total treatment time was 60 s) at a pressure of 50 mTorr, where the reactive gas was 5% of the total flow rate [55]. The ion energy was on the average 3 eV, which the latter is lower than the bonding energy of C-C (3.6 eV) or C=C bonds (6.35 eV) that form the graphitic plane of graphene. The total fluorine content after plasma treatment, measured by the XPS method, was 18 at.%. Moreover, no sulfur or sulfur compounds were observed after treatment. The intensity of the D band in Raman spectra caused by the disturbances increases significantly and exceeds the intensity G of the peak. Although the incident ion energies are low to cause impact-driven physical defects, the ion energies are sufficient to drive chemical reactions at the surface when reactive ions and/or neutrals are present during irradiation. Fluorine forms two types of bonds with carbon: ionic and semi-ionic bonds, which do not violate the planar character of graphene, and covalent sp^3-hybridized bonds. The increase in the intensity of the D band in the Raman spectra indicates an increasing amount of carbon bound to sp^3 in the structure of the modified grapheme [18, 33]. After heat treatment at T = 500°C in an Ar atmosphere with a duration of 10 min, the fluorine is completely removed, the intensities of D and D′ peaks, as well as the I_D/I_G ratio drastically decrease [55]. In [56] the effect of pure SF_6 plasma generated in an reactive ion etcher (RIE) system at a radio-frequency of 13.56 MHz on epitaxial grapheme. An rf power of 50 W and an SF_6 partial pressure of 100 mTorr were used for all experiments. From XPS, ultraviolet photoelectron spectroscopy (UPS) and Raman spectroscopy of multilayer epitaxial graphene, the authors concluded that the configuration of sp^2 graphene remains unchanged after plasma treatment, and fluorination is limited to one or two surface layers. The authors believe that fluorination to the carbon atoms at the edges of graphite domains generated by ion-bombardment. Similar studies of the effect on graphene obtained by a micromechanical method were carried out by Yang [57]. The plasma processing conditions were as follows: 5 Pa of pressure, 5 W of power and a 5 sccm gas feed rate by different durations. From Raman spectra studies, the authors report that fluorination of single-layer graphene occurs is much more feasible than that of bilayer and trilayer graphenes, for which high fluorination times are required. Studies have shown that the process of plasma fluorination of graphene is a reversible process. The annealing carried out by the authors at a temperature of 970° K for 1 hour completely restored the original graphene structures.

Chen et al. [58] from the Raman analysis of spectra of graphene exposed in plasma SF_6 (5 Pa, 5 W, 5 sccm) concluded about p-doping of graphene. Moreover, the p-doping of the edges of monolayer graphene is more significant than its central part.

Zhang [59] used SF_6 plasma to fluorinate the CVD graphene (37.5 mTorr, P = 5 W, at a gas flow rate of 2 sccm, DC bias voltage 13 V, the processing time was from 10 to 90 s). From the analysis of Raman spectra and XPS, the authors concluded that fluorination leads to covalent bonds and p-doping of graphene occurs. At the same time, the ratio I_D/I_G, which increased to ~ 2.9 after initial treatments of up to 50 s, starts to decrease continuously with further increase in the processing time. The authors explain this effect by the formation of a less stable fluorine group, which decays with increasing processing time.

The results of the plasma fluorination of graphene in CF_4 are reported in Refs. [26, 60–66]. Cheng et al. used the RIE system to treat CVD graphene CF_4 with plasma (40 W, 700 mTorr) [67]. As a result of plasma processing in Raman spectra of graphene, after 10 s of processing, the intensities of the peaks (D, D′, D + G), associated with the introduction of structural lattice disorderings [18, 33, 62], significantly increased. According to the authors, the reason for this can be the conversion of sp^2-carbon to sp^3-hybridization due to the adsorption of fluorine, which coincides with the conclusions of other authors [61, 62]. With an increase in processing time to 300 s, Raman peaks become almost invisible. This result is typical of strongly fluorinated graphene [68–70]. It was shown in report [60] that the ratio of the intensity D of the peak to the G peak increases significantly with initial treatments and then goes to saturation with a further increase in the fluorination time. At the same time, the ratio of I_{2D}/I_G intensities shows the opposite trend and reaches to saturation gradually. The intensity of the 2D Raman peak is related to two phonon double-resonance Raman processes [33, 71]. The saturation yield can mean the absence of chemical etching by fluorine of carbon, if there is no ion bombardment [62]. At the same time, Shen and other [63] methods of optical microscopy and Raman spectroscopy observed thinning of graphene layers and an increase in structural disturbances during fluorination in CF_4 plasma (at a power of 20 W at a pressure of 0.8 Torr), as well as in the report [52]. Exposure of graphene in a plasma with a duration of 5 s resulted in the removal of the upper layer of graphene. The paper notes that functionalization occurs due to the formation of covalent bonds that distort the lattice structures of graphene. As a result, the intensities of D and D ′peaks in the Raman spectra increase. It should be noted that different authors have no common opinion on this matter. From the analysis of XPS data and the results of measuring electrical characteristics, Cheng et al. in [60] state that at low fluorine content, ionic bonds of C-F components are introduced. With a high content of fluorine, covalent bonds dominate. The above increase in I_D/I_G and a decrease in the I_{2D}/I_G ratio in the Raman spectra, with short processing times (up to 10 s), was observed in [26, 61, 62] under close fluorination conditions in plasma. The Raman I_D/I_G peak ratio mapping images of fluorinated graphene showed that the flat portions of graphene are uniformly fluorinated [60]. While, in multilayered CVD graphene containing wrinkles, wrinkles, etc., heterogeneous fluorination occurs. As the authors argue, these areas are less susceptible to fluoridation. Similar results were obtained in [26, 61], who found that single-layer graphene is more efficiently fluorinated by plasma than two and three-layer graphene films. Measurements of the resistance of CF_4 CVD-graphene fluorinated in plasma showed a significant increase in electrical resistance from several kΩ to several MΩ, [65] and even more than 100 GΩ [64]. Despite the high values of electrical resistance, there were regions with low resistances (such as bilayer islands, folds (2-layer height, width ~ 100 nm), wrinkles (line width < 50 nm), and ripples (fine parallel lines with spacing ~ 150 nm), which have small resistance [64]. In the authors′ opinion, this is due to the weak fluorination of these regions. The increase in the resistance of fluorinated graphene is associated with the formation of covalent bonds [64, 65]. Annealing in a 30-min nitrogen atmosphere showed that the fluoride desorption process is heterogeneous, due to the presence of various C-F bonding components, and can begin at temperatures of 150°C with a low fluorine content [60]. Annealing of samples with a high content of fluorine, which has predominantly covalent components, showed that at T = 300°C, the ratio of I_D/I_G intensities increases to 1.8, and then decreases to ~ 1.5 when annealed to 600°C [60]. In [62], from the measurements of topography and currents obtained by the AFM method, fluorinated clusters sp3, with dimensions of ~ 20–30 nm, were detected.

An increase in the covalent bound fluorine content during plasma fluorination was observed on the surface of graphene nanoplates (GNPs), multi-wall carbon nanotubes (MWCNT) [72, 73] and highly ordered pyrolytic graphite (HOPG) [72]. Fluorination of GNPs and MWCNT in CF_4 plasma leads to covalent bonding of fluorine with irregularities, defects, boundaries and with sp^2 carbon atoms inside graphene sheets of both carbon materials [72]. Thus, the surface area of both materials is a parameter that partially determines the degree of functionalization. This is confirmed by the fact that surfaces of CNTs are easier to fluorinate than HOPG surfaces. This is the expected result, since the hybridized compound sp^2 in graphite is much more stable than the distorted sp^2 compound observed in nanotubes [21].

In reports of Tahara et al. [74] to suppress ion bombardments and improve the reaction with fluorine radicals on graphene, the substrate was placed "face down" in the plasma chamber. Graphene samples were investigated prepared by the mechanical exfoliation from graphite. For the fluorination process was used a RIE system. The graphene samples were exposed to Ar/F2 (90%/10%) plasma with a relatively low rf power of 5 W, a gas pressure of 0.1 Torr, and a total gas flow rate of 75 sccm at room temperature. The reaction time ranged from 0.5 to 30 min. The fluorination of graphene led to a sharp increase in the intensity of D-peak caused by the defects (1350 cm^{-1}). At the same time, with increasing reaction time, the D peak intensity shown non-monotonic behavior and had the maximum at 3 min, while the 2D peak intensity decreased monotonously. As the authors believe, such behavior of the peak intensities indicates an increase of amount of fluorine atoms attaching to the graphene. The authors explain the decrease in the D-peak intensity by the competition between the defect-phonon scattering processes and the decrease in the lifetime of the electronic states, caused by an increase in the defect concentration. The paper also shows that the monolayer graphene was more reactive than bilayer. In addition, an annealed the fluorinated graphene samples in the atmosphere for 90 min at 573 K showed that fluorination in a plasma is a reversible process.

5. Conclusion

A review of the literature shows that plasma treatments in various gases are primarily used for the functionalization of graphene oxide to produce graphene oxide from graphene. Summarizing the effects of GO processing in various gases, the following conclusions can be drawn:

1. The effect of an oxygen-containing plasma leads to rapid etching of the GO layers, accompanied by the formation of a large number of defects. The etching rates depend on the type of plasma source used, the location of the samples in the reaction chamber and the distance

from the plasma, as well as the plasma power, gas pressure and processing temperature. The aggressive effect of oxygen plasma can be reduced by using 'gentle' treatments. In addition, plasma treatment in oxygen can be used to produce GO from graphene. The graphene oxide obtained in this way has a higher wettability of the surface, this circumstance is of interest in the development of biological sensors and gas sensors.

2. Treatment in nitrogen plasma does not lead to aggressive etchings in oxygen, but if special measures are not taken to protect the exhaust gas from the plasma, defect formation can be significant. Exposure of graphene oxide in a nitrogen plasma can be used for n-doping graphene oxide, but p-doping has been observed in some studies. This effect depends on the type of binding of nitrogen atoms to carbon (formation of pyridine, pyrrole and graphite configurations) and, depending on this, both donor and acceptor charge carriers can be introduced. The use of ammonia NH_3 plasma allows for both reduction and simultaneous alloying of graphene oxide.

3. Plasma treatment in fluorine-containing plasma (CF_4, SF_6) is an effective method of fluorination of GO, mainly due to the displacement of oxygen atoms. The intensity of etching in the fluorine plasma is low, which allows the use of treatments lasting up to tens of minutes. The effect of plasma leads to an increase in the electrical resistance, accompanied by a weak p-doping. Plasma fluorination is a reversible effect and when heat treatments above 500°C, complete removal of fluorine atoms occurs.

The general effects of the action of plasma treatment can be reduced to the following points: the most noticeable functionalization of graphene oxide occurs at the initial processing times in plasma, duration, as a rule, up to 1 min. The effect of plasma on GO properties is limited to several atomic layers and does not affect the bulk properties of GO. When plasma treatments GO one of the primary tasks remains to protect the surface from the defect formation and etching of GO films.

In general, it should be noted that the effect of plasma treatments on GO properties is still poorly understood. If we compare the amount of work devoted to the plasma treatment of graphene and graphene oxide, then this amount for GO is much lower. A still poorly studied region remains the study of the effect of a plasma of a mixture of various gases on the properties of graphene oxide. Mechanisms of functionalization of graphene oxide in various plasma media have not been fully investigated. The problem of the "gentle" effect of plasma on the surface of graphene oxide has not been completely solved. When using plasma treatments to control the properties of graphene oxide, these problems must be solved in the future.

Author details

Efim Petrovich Neustroev

Address all correspondence to: neustr@mail.ru

Institute of Physics and Technologies, North-Eastern Federal University (NEFU), Yakutsk, Russia

References

[1] Gao W, editor. Graphene oxide. In: Reduction, Spectroscopy, and Applications. Switzerland: Springer International Publishing; 2015. p. 147. DOI: 10.1007/978-3-319-15500-5

[2] Wang X, Shi G. Introduction to the chemistry of graphene. Physical Chemistry Chemical Physics. 2015;**17**:28484. DOI: 10.1039/C5CP05212B

[3] Alam SN, Sharma N, Kumar L. Synthesis of graphene oxide (GO) by modified hummers method and its thermal reduction to obtain reduced graphene oxide (rGO). Graphene. 2017;**6**:1-18. DOI: 10.4236/graphene.2017.61001

[4] Chen D, Feng H, Li J. Graphene oxide: Preparation, functionalization, and electrochemical applications. Chemical Reviews. 2012;**112**(11):6027-6053. DOI: 10.1021/cr300115g

[5] Ray SC, Ray SC. Applications of Graphene and Graphene-Oxide Based Nanomaterials. Switzerland: Springer International Publishing; 2015. p. 84. DOI: 10.1016/B978-0-323-37521-4.00002-9

[6] Zhao H, Fan S, Chen Y, Feng Z, Zhang H, Pang W, Zhang D, Zhang M. Graphene oxide surface functionalization for sensitivity enhancement of thin-film piezoelectric acoustic gas sensors. ACS Applied Materials & Interfaces. 2017;**9**(46):40774-40781. DOI: 10.1021/acsami.7b09547

[7] Hafiz SM, Ritikos R, Whitcher TJ, Razib N, Bien DCS, Chanlek N, Nakajima H, Saisopa T, Songsiririttthigul P, Huang NM, Rahman SA. A practical carbon dioxide gas sensor using room-temperature hydrogen plasma reduced graphene oxide. Sensors and Actuators B: Chemical. 2014;**193**:692-700. DOI: 10.1016/j.snb.2013.12.017

[8] Yu Y, Kang BH, Lee YD, Lee SB, Ju BK. Effect of fluorine plasma treatment with chemically reduced graphene oxide thin films as hole transport layer in organic solar cells. Applied Surface Science. 2013;**287**:91-96. DOI: 10.1016/j.apsusc.2013.09.078

[9] Chae MS, Kim J, Jeong D, Kim Y, Roh JH, Lee SM, Heo Y, Kang JY, Lee JH, Yoon DS, Kim TG, Chang TS, Hwang KS. Enhancing surface functionality of reduced graphene oxide biosensors by oxygen plasma treatment for Alzheimer's disease diagnosis. Biosensors and Bioelectronics. 2017;**92**:610-617. DOI: 10.1016/j.bios.2016.10.049

[10] Wang Y, Shao Y, Matson DW, Li J, Lin Y. Nitrogen-doped graphene and its application in electrochemical biosensing. AcsNano. 2010;**4**(4):1790-1798. DOI: 10.1021/nn100315s

[11] Ke Z, Ma Y, Zhu Z, Zhao H, Wang Q, Huang Q. Non-thermal hydrogen plasma processing effectively increases the antibacterial activity of graphene oxide. Applied Physics Letters. 2018;**112**(1):013701. DOI: 10.1063/1.5012132

[12] Lee SW, Mattevi C, Chhowalla M, Sankaran RM. Plasma-assisted reduction of graphene oxide at low temperature and atmospheric pressure for flexible conductor applications. Journal of Physical Chemistry Letters. 2012;**3**(6):772-777. DOI: 10.1021/jz300080p

[13] Huang CH, Wang YY, Lu TH, Li YC. Flexible transparent electrode of hybrid Ag-nanowire/reduced-graphene-oxide thin film on PET substrate prepared using H_2/Ar low-damage plasma. Polymer. 2017;**9**(1):28. DOI: 10.3390/polym9010028

[14] Walton SG, Hernández SC, Boris DR, Petrova TB, Petrov GM. Electron beam generated plasmas for the processing of graphene. Journal of Physics D: Applied Physics. 2017;**50**: 354001. DOI: 10.1088/1361-6463/aa7d12

[15] Dey A, Chroneos A, Braithwaite NSJ, Gandhiraman RP, Krishnamurthy S. Plasma engineering of graphene. Applied Physics Reviews. 2016;**3**:021301. DOI: 10.1063/1.4947188

[16] Seah CM, Vigolo B, Chai SP, Mohamed AR. Mechanisms of graphene fabrication through plasma-induced layer-by-layer thinning. Carbon. 2016;**105**:496e509. DOI: 10.1016/j.carbon.2016.04.072

[17] Bianco GV, Sacchetti A, Ingrosso C, Giangregorio MM, Losurdo M, Capezzuto P, Bruno G. Engineering graphene properties by modulated plasma treatments. Carbon. 2018;**129**:869-877. DOI: 10.1016/j.carbon.2017.11.015

[18] Childres I, Jauregui LA, Tian J, Chen YP. Effect of oxygen plasma etching on graphene studied with Raman spectroscopy and electronic transport. New Journal of Physics. 2011; **13**:025008. DOI: 10.1088/1367-2630/13/2/025008

[19] Rao F, Li W, Dong L. Layer engineering of graphene with oxygen plasma etching. In: Proceding of the 11th IEEE International Conference on Nanotechnology; 15–18 August 2011; Portland. Oregon: IEEE; 2011. pp. 1201-1204

[20] Kondratowicz I, Nadolska M, Şahin S, Łapiński M, Prześniak-Welenc M, Sawczak M, Yu EH, Sadowski W, Żelechowska K. Tailoring properties of reduced graphene oxide by oxygen plasma treatment. Applied Surface Science. 2018;**440**:651-659. DOI: 10.1016/j.apsusc.2018.01.168/

[21] Cheng HE, Wang YY, Wu PC, Huang CH. Preparation of large-area graphene oxide sheets with a high density of carboxyl groups using O_2/H_2 low-damage plasma. Surface and Coatings Technology. 2016;**303**:170-175. DOI: 10.1016/j.surfcoat.2016.03.028

[22] Huang CH, Su CY, Lai CS, Li YC, Samukawa S. Ultra-low-damage radical treatment for the highly controllable oxidation of large-scale graphene sheets. Carbon. 2014;**73**:244-251. DOI: 10.1016/j.carbon.2014.02.060

[23] Ostovari F, Abdi Y, Ghasemi F. Controllable formation of graphene and graphene oxide sheets using photo-catalytic reduction and oxygen plasma treatment. European Physical Journal Applied Physics. 2012;**60**:30401. DOI: 10.1051/epjap/2012120338

[24] Choi K, Lim J, Rani JR, Yoon HS, Oh J, Hong T, Ha T, Park BC, Sim KI, Jun SC, Kim JH. Terahertz and optical study of monolayer graphene processed by plasma oxidation. Applied Physics Letters. 2013;**102**:131901. DOI: 10.1063/1.4795526

[25] Lee BJ, Jeong GH. Plasma oxidation of thermally grown graphenes and their characterization. Vacuum. 2013;**87**:200e204. DOI: 10.1016/j.vacuum.2012.02.022

[26] Felten A, Eckmann A, Pireaux JJ, Krupke R, Casiraghi C. Controlled modification of mono- and bilayer graphene in O_2, H_2 and CF_4 plasmas. Nanotechnology. 2013;**24**:355705. DOI: 10.1088/0957-4484/24/35/355705

[27] McEvoy N, Nolan H, Nanjundan AK, Hallam T, Duesberg GS. Functionalization of graphene surfaces with downstream plasma treatments. Carbon. 2013;**54**:283-290. DOI: 10.1016/j.carbon.2012.11.040

[28] Lim WS, Kim YY, Kim H, Jang S, Kwon N, Park BJ, Ahn JH, Chung I, Hong BH, Yeom GY. Atomic layer etching of graphene for full graphene device fabrication. Carbon. 2012;**50**: 429-435. DOI: 10.1016/j.carbon.2011.08.058

[29] Santoso I, Singh RS, Gogoi PK, Asmara TC, Wei D, Chen W, Wee ATS, Pereira VM, Rusydi A. Tunable optical absorption and interactions in graphene via oxygen plasma. Physical Review B. 2014;**89**:075134. DOI: 10.1103/PhysRevB.89.075134

[30] Kim JH, Ko E, Hwang J, Pham XH, Lee JH, Lee SH, Tran VK, Kim JH, Park J, Choo J, Han KN, Seong GH. Large scale plasma patterning of transparent graphene electrode on flexible substrates. Langmuir. 2015;**31**(9):2914-2921. DOI: 10.1021/la504443a

[31] Pei S, Cheng HM. The reduction of graphene oxide. Carbon. 2012;**50**:3210-3228. DOI: 10.1016/j.carbon.2011.11.010

[32] Lavanya J, Gomathi N, Neog S. Electrochemical performance of nitrogen and oxygen radio frequency plasma induced functional groups on tri layered reduced graphene oxide. Materials Research Express. 2014;**1**(2):025604. DOI: 10.1088/2053-1591/1/2/025604

[33] Beams R, Cancado LG, Novotny L. Raman characterization of defects and dopants in graphene. Journal of Physics: Condensed Matter. 2015;**27**:083002. DOI: 10.1088/0953-8984/27/8/083002

[34] Mao H, Wang R, Zhong J, Zhong S, Chen W. Mildly O_2 plasma treated CVD graphene as a promising platform for molecular sensing. Carbon. 2014;**76**:212-219. DOI: 10.1016/j.carbon.2014.04.070

[35] Li Z, Xu Y, Cao B, Qi L, He S, Wang C, Zhang J, Wang J, Xu K. Raman spectra investigation of the defects of chemical vapor deposited multilayer graphene and modified by oxygen plasma treatment. Superlattices and Microstructures. 2016;**99**:125e130. DOI: 10.1016/j.spmi.2016.03.035

[36] Gokus T, Nair RR, Bonetti A, Böhmler M, Lombardo A, Novoselov KS, Geim AK, Ferrari AC, Hartschuh A. Making graphene luminescent by oxygen plasma treatment. ACS Nano. 2009;**3**(12):3963-3968. DOI: 10.1021/nn9012753

[37] Ucar N, Can E, Yuksek İO, Olmez M, Onen A, Yavuz NK. The effect of exfoliation and plasma application on the properties of continuous graphene oxide fiber. Fullerenes, Nanotubes, and Carbon Nanostructures. 2017;**25**(10):570-575. DOI: 10.1080/1536383X.2017.1337749

[38] Kim HT, Kim C, Park C. Reduction and nitridation of graphene oxide (GO) films at room temperature using inductively coupled NH_3 plasma. Vacuum. 2014;**108**:35-38. DOI: 10.1016/j.vacuum.2014.05.018

[39] Kim MJ, Jeong Y, Sohn SH, Lee SY, Kim YJ, Lee K, Kahng YH, Jang JH. Fast and low-temperature reduction of graphene oxide films using ammonia plasma. AIP Advances. 2013;**3**:012117. DOI: 10.1063/1.4789545

[40] Lee SY, Kim C, Kim HT. Difference in chemical reactions in bulk plasma and sheath regions during surface modification of graphene oxide film using capacitively coupled NH_3 plasma. Journal of Applied Physics. 2015;**118**:103303. DOI: 10.1063/1.4930044

[41] Kumar NK, Nolan H, McEvoy N, Rezvani E, Doyle RL, Lyons MEG, Duesberg GS. Plasma-assisted simultaneous reduction and nitrogen doping of graphene oxide nanosheets. Journal of Materials Chemistry A. 2013;**1**:4431-4435. DOI: 10.1039/c3ta10337d

[42] Mohai M, Bertóti I. Modification of graphene-oxide surface in nitrogen and argon glow discharge plasma. Surface and Interface Analysis. 2016;**48**(7):461-464. DOI: 10.1002/sia.5929

[43] Neustroev EP, Prokopiev AR, Soloviev BD, Burtseva EK, Popov VI, Timofeev VB. Modification of graphene oxide films by radiofrequency N_2 plasma. Nanotechnology, Focus on Electronics, Photonics and Renewable Energy. 2018;**29**(14):144002. DOI: 10.1088/1361-6528/aaabe3

[44] Usachov D, Vilkov O, Grüneis A, Haberer D, Fedorov A, Adamchuk VK, Preobrajenski AB, Dudin P, Barinov, Oehzelt M, Laubschat C, Vyalikh DV. Nitrogen-doped graphene: Efficient growth, structure, and electronic properties. NanoLetters. 2011;**11**:5401-5407. DOI: 10.1021/nl2031037

[45] Rybin M, Pereaslavtsev A, Vasilieva T, Myasnikov V, Sokolov I, Pavlova A, Obraztsova E, Khomich A, Ralchenko V, Obraztsova E. Efficient nitrogen doping of graphene by plasma treatment. Carbon. 2016;**96**:196-202. DOI: 10.1016/j.carbon.2015.09.056

[46] Lin YP, Ksari Y, Prakash J, Giovanelli L, Valmalette JC, Themlin JM. Nitrogen-doping processes of graphene by a versatile plasma-based method. Carbon. 2014;**73**:216-224. DOI: 10.1016/j.carbon.2014.02.057

[47] Shao Y, Zhang S, Engelhard MH, Li G, Shao G, Wang Y, Liu J, Aksayc IA, Lin Y. Nitrogen-doped graphene and its electrochemical applications. Journal of Materials Chemistry. 2010;**20**:7491-7496. DOI: 10.1039/c0jm00782j

[48] Tao H, Yan C, Robertson AW, Gao Y, Ding J, Zhang Y, Ma T, Sun Z. N-doping of graphene oxide at low temperature for oxygen reduction reaction. Chemical Communications. 2017; **53**:873-876. DOI: 10.1039/x0xx00000x

[49] Qin S, Xu Q. Room temperature ferromagnetism in N_2 plasma treated grapheme oxide. Journal of Allows and Compounds. 2016;**692**:332-338. DOI: 10.1016/j.jallcom.2016.09.055

[50] Wang Y, Yu F, Zhu M, Ma C, Zhao D, Wang C, Zhou A, Dai B, Ji J, Guo X. N-doping of plasma exfoliated graphene oxide via dielectric barrier discharge plasma treatment for oxygen reduction reaction. Journal of Materials Chemistry A. 2018;**6**:2011-2017. DOI: 10.1039/x0xx00000x

[51] Zhu H, Ji D, Jiang L, Dong H, Hu W. Tuning electrical properties of graphite oxide by plasma. Philosophical Transactions of the Royal Society A. 2013;**371**:20120308. DOI: 10.1098/rsta.2012.0308

[52] Zhou B, Qian X, Li M, Ma J, Liu L, Hu C, Xu Z, Jiao X. Tailoring the chemical composition and dispersion behavior of fluorinated graphene oxide via CF_4 plasma. Journal of Nanoparticle Research. 2015;**17**:130. DOI: 10.1007/s11051-015-2946-0

[53] Neustroev EP, Nogovitcyna MV, Popov VI, Timofeev VB. Modification of thermally reduced graphene oxide by the SF_6/Ar plasma treatment. Inorganic Materials: Applied Research. 2017;**8**(5):763-768. DOI: 10.1134/S2075113317050215

[54] Yu X, Lin K, Qiu K, Cai H, Li X, Liu J, Pan N, Fu S, Luo Y, Wang X. Increased chemical enhancement of Raman spectra for molecules adsorbed on fluorinated reduced graphene oxide. Carbon. 2012;**50**:4512-4517. DOI: 10.1016/j.carbon.2012.05.033

[55] Baraket M, Walton SG, Lock EH, Robinson JT, Perkins FK. The functionalization of graphene using electron-beam generated plasmas. Applied Physics Letters. 2010;**96**. DOI: 231501, 10.1063/1.3436556

[56] Sherpa SD, Paniagua SA, Levitin G, Marder SR, Williams MD, Hess DW. Photoelectron spectroscopy studies of plasma-fluorinated epitaxial graphene. Journal of Vacuum Science and Technology. 2012;**B30**:03D102. DOI: 10.1116/1.3688760

[57] Yang H, Chen M, Zhou H, Qiu C, Hu L, Yu F, Chu W, Sun S, Sun L. Preferential and reversible fluorination of monolayer graphene. Journal of Physical Chemistry C. 2011;**115**: 1684416848. DOI: 10.1021/jp204573z

[58] Chen M, Qiu C, Zhou H, Yang H, Yu F, Sun L. Fluorination of edges and central areas of monolayer graphene by SF_6 and CHF_3 plasma treatments. Journal of Nanoscience and Nanotechnology. 2013;**13**:1331-1334. DOI: 10.1166/jnn.2013.5996

[59] Zhang H, Fan L, Dong H, Zhang P, Nie K, Zhong J, Li Y, Guo J, Sun X. Spectroscopic investigation of plasma-fluorinated monolayer graphene and application for gas sensing. ACS Applied Materials & Interfaces. 2016;**8**(13):8652-8661. DOI: 10.1021/acsami.5b11872

[60] Cheng L, Jandhyala S, Mordi G, Lucero AT, Huang J, Azcatl A, Addou R, Wallace RM, Colombo L, Kim J. Partially fluorinated graphene: Structural and electrical characterization. ACS Applied Materials & Interfaces. 2016;**8**(7):5002-5008. DOI: 10.1021/acsami.5b11701

[61] Chen M, Zhou H, Qiu C, Yang H, Yu F, Sun L. Layer-dependent fluorination and doping of graphene via plasma treatment. Nanotechnology. 2012;**23**:115706. DOI: 10.1088/0957-4484/23/11/115706

[62] Eckmann A, Felten A, Mishchenko A, Britnell L, Krupke R, Novoselov KS, Casiraghi C. Probing the nature of defects in graphene by Raman spectroscopy. Nano Letters. 2012; **12**(8):3925-3930. DOI: 10.1021/nl300901a

[63] Shen C, Huang G, Cheng Y, Cao R, Ding F, Schwingenschlögl U, Mei Y. Thinning and functionalization offew-layer graphene sheets by CF_4 plasma treatment. Nanoscale Research Letters. 2012;**7**:268. DOI: 10.1186/1556-276X-7-268

[64] Wang B, Wang J, Zhu J. Fluorination of graphene: A spectroscopic and microscopic study. AcsNano. 2014;**8**(2):1862-1870. DOI: 10.1021/nn406333f

[65] Ho KI, Liao JH, Huang CH, Hsu CL, Zhang W, AY L, Li LJ, Lai CS, Su CY. One-step formation of a single atomic-layer transistor by the selective fluorination of a graphene film. Small. 2014;**10**(5):989-997. DOI: 0.1002/smll.201301366

[66] Bon SB, Valentini L, Verdejo R, Fierro JLG, Peponi L, Lopez-Manchado MA, Kenny JM. Plasma fluorination of chemically derived graphene sheets and subsequent modification with butylamine. Chemistry of Materials. 2009;**21**:3433-3438. DOI: 10.1021/cm901039j

[67] Feng W, Long P, Feng Y, Li Y. Two-dimensional fluorinated graphene: Synthesis, structures, properties and applications. Advancement of Science. 2016;**3**:1500413. DOI: 10.1002/advs.201500413

[68] Nair RR, Ren W, Jalil R, Riaz I, Kravets VG, Britnell L, Blake P, Schedin F, Mayorov AS, Yuan S, Katsnelson MI, Cheng HM, Strupinski W, Bulusheva LG, Okotrub AV, Grigorieva IV, Grigorenko AN, Novoselov KS, Geim AK. Fluorographene: A two dimensional counterpart of teflon. Small. 2010;**6**(24):2877-2884. DOI: 10.1002/smll.201001555

[69] Wang X, Dai Y, Gao J, Huang J, Li B, Fan C, Yang J, Liu X. High-yield production of highly fluorinated graphene by direct heating fluorination of graphene-oxide. ACS Applied Materials & Interfaces. 2013;**5**(17):8294-8309. DOI: 10.1021/am402958p

[70] Nebogatikova NA, Antonova IV, Prinz VY, Volodin VA, Zatsepin DA, Kurmaev EZ, Zhidkov IS, Cholakh SO. Functionalization of graphene and few-layer graphene films in an hydrofluoric acid aqueous solution. Nanotechnology in Russia. 2014;**9**(1–2):51. DOI: 10.1134/S1995078014010108

[71] Ferrari AC, Meyer JC, Scardaci V, Casiraghi C, Lazzeri M, Mauri F, Piscanec S, Jiang D, Novoselov KS, Roth S, Geim AK. Raman spectrum of graphene and graphene layers. Physical Review Letters. 2006;**97**(18):187401. DOI: 10.1103/PhysRevLett.97.187401

[72] Abdelkader-Fernándeza VK, Morales-Laraa F, Melguizob M, García-Gallarínb C, López-Garzónb R, Godino-Salidob ML, López-Garzóna FJ, Domingo-Garcíaa M, Pérez-Mendozaa MJ. Degree of functionalization and stability of fluorine groups fixed to carbon nanotubes and graphite nanoplates by CF_4 microwave plasma. Applied Surface Science. 2015;**357**:1410-1418. DOI: 10.1016/j.apsusc.2015.09.262

[73] Barlow A, Birch A, Deslandes A, Quinton JS. Plasma fluorination of highly ordered Pyrolytic graphite and single walled carbon nanotube surfaces. In: Proceedings of the International Conference on Nanoscience and Nanotechnology (ICONN '06), 3–7 July 2006. Brisbane: IEEE; 2006. pp. 103-106. DOI: 10.1109/ICONN.2006.340561

[74] Tahara K, Iwasaki T, Matsutani A, Hatano M. Effect of radical fluorination on mono- and bi-layer graphene in Ar/F_2 plasma. Applied Physics Letters. 2012;**01**:163105. DOI: 10.1063/1.4760268

Chapter 3

Carbocatalytic Activity of Graphene Oxide in Organic Synthesis

Biju Majumdar, Daisy Sarma and Tridib K. Sarma

Additional information is available at the end of the chapter

http://dx.doi.org/10.5772/intechopen.77361

Abstract

Nanochemistry has evolved as an important part in catalysis for both academic as well as industrial research. Traditional homogeneous catalytic systems have gained significant importance due to the molecular level analysis of their catalytic activity and the excellent homogeneity of the catalysts and the reactants. However, removal of the catalysts from the reaction mixture without product contamination requires tedious purification steps. With increasing ecological and economical demands towards sustainable chemical synthesis, the recovery and reuse of catalysts has been an important factor. In this drive, various heterogeneous catalytic systems including mesoporous materials, solid catalysts, organometallics, noble-metal nanoparticles, etc. have been developed for photochemical and electrochemical conversion, environmental remediation as well as catalyst for important chemical transformations. Carbon nanomaterial specially graphene oxide and carbon dots have received significant research importance due to their large scale availability, easy surface modification, non-toxicity and other surface properties. Here, we review the continuous progress in the development of graphene based materials and their catalytic activity in organic synthesis.

Keywords: carbocatalysis, graphene oxide, oxygen functional groups, organic synthesis

1. Introduction

In recent years, there has been tremendous focus towards developing greener synthetic methods for the industrial production of fine and commodity chemicals. Towards the development of economical and sustainable routes for large scale synthesis, the ideal protocols are characterized by four parameters: catalytic activity, selectivity, atom-economy and step-selectivity. Among them, development of catalytic systems with high activity and selectivity plays the most important role. A catalyst provides an alternative pathway with a lower activation energy barrier for an organic reaction without being transformed itself [1]. If there are several products and several reaction barriers, a catalyst can drive the reaction selectivity with a particular product. The efficiency of a catalyst lies on the formation of the desired product with high yield and selectivity at relatively milder reaction conditions. Hence, the search for suitable catalytic systems to achieve green and sustainable production of chemicals is growing continuously, that promotes reduction of toxic and hazardous chemicals, stoichiometric amount of promoter, less consumption of energy, fewer side products and less number of steps involved.

In the past, the main objective of catalysis was to enhance the activity and selectivity of a catalyst, the recovery and reusability of a catalyst was rarely a major concern. Traditional homogeneous organocatalysts, transition metal catalysts as well as bio-catalytic systems are highly efficient because the catalytic activity can be defined on a molecular level. These single site catalysts are highly accessible to reactant molecules, readily soluble in reaction medium and often give rise to high catalytic activity and selectivity even under mild conditions [2]. However, often use of expensive metal salts, expensive ligands, expensive and tedious purification steps as well as recovery of the catalyst is the major issue. A variety of homogeneous and heterogeneous catalysts have been developed, however there is still a vast scope for development of suitable catalysts that not only offers high activity and selectivity, but also provide a greener route. Fabricating single step synthetic methods using heterogeneous catalysts is a challenging goal as it aims at decreasing the energy and time consumption by operation of single step synthesis. This consequently eliminates the need of separation of products from the reaction mixture again and again.

1.1. Nanocatalysis

In the area of heterogeneous catalysis, the primary focus is towards development of materials with high surface area as the reactions involve surface initiated pathway. With exceptional nanodimensional properties (high surface to volume ratio), nanoparticles have produced tremendous interest in wide range of research activities including pharmaceuticals, fine chemicals, renewable energy and biotransformation. The key objective of nanocatalysis research is to produce catalysts with 100% selectivity with extremely high activity, low energy consumption, and long lifetime. This can be achieved only by precisely controlling the size, shape, spatial distribution, surface composition and electronic structure, and thermal and chemical stability of the individual nanocomponents. In addition, surface atoms which are at the edges or in the corners are more active than those in planes, and their number also increases with decreasing particle size. Thus nanoparticles have shown tremendous applicability at the interface between homogeneous and heterogeneous catalysis [3–5]. Homogeneous nanocatalysts are used in the same medium as the reactants. The main concern with homogeneous nanocatalysts is their recovery from the solution for repeated cycles as nanoparticles are extremely difficult to be removed from a solution and the extra steps needed to do so could completely negate the process simplification. If the nanoparticles cannot be recovered, they pose an environmental risk, as well as threatening the profitability of the process. For easy recovery and recyclability, the active nanocatalysts are often immobilized on a solid inert matrix, which is in a different phase to the reactants. The solid catalysts can in most cases simply be filtered out and used for the next cycle of reactions, making the process economical and ecological. Due to the complex physicochemical properties at the nanometer level and possibility of multiple surface initiated reactions at the active site, synthesis-structure-catalysis performance relationships are poorly understood in many catalytic systems. A variety of heterogeneous catalytic systems based on metal nanoparticles, semiconductors, oxides/sulfides, mesoporous materials, metal-organic framework etc. and their combinations have been developed for photochemical and electrochemical catalysis, environmental remediation as well as catalysis for important chemical conversions. Other metal-free nanomaterials such as fullerene, graphite, graphene oxide (GO), carbon nanotube (CNT) and carbon nanodots (CNDs) have also been studied either as effective carbocatalyst or as support for metal/oxide for various catalytic applications.

In heterogeneous catalysis, the modus operandi is through surface mediated reactions. Therefore, availability of large surface area is a prerequisite for high catalytic conversions [6, 7]. Although, unsupported nanoparticles with well-defined surface structure and clean exposed facets are predicted to be highly active for catalysis as shown by several theoretical studies, their applicability in real catalysis is a challenge. Due to high surface energy resulting from the large fraction of atoms present on the surface of NPs, they tend to reduce the surface energy and stabilize themselves through agglomeration and coagulation, resulting in decreased active surface area [7]. This leads to an exponential decrease in the surface area for effective catalysis and the prime objective of using nanoparticles for repeated cycles with high activity is practically lost. For example, in case of Au nanoparticles, size plays a critical role during oxidation reactions, as nanoparticles with the diameter 1–5 nm are highly active, whereas larger nanoparticles becomes inactive. The protection of nanoparticle surface from agglomeration can be achieved by using a surface stabilizing agent or immobilizing the NPs into a solid matrix with high surface areas [8, 9]. Soft organic materials such as polymers, surfactants, dendrimers, and ionic liquids have been used as effective capping agents. These systems often show high catalytic performances [10] e.g. reduction of nitrobenzene, olefin hydrogenation and CO oxidation at low temperatures by Au nanoparticles, C—C coupling or hydrosilylation reactions of olefins by Pd nanoparticles. However, these polymeric supports suffer certain disadvantages such as providing low surface area which limits the interface between the catalyst and the reaction substrate. Moreover, soft nature of the materials and low mechanical stability often leads to separation of the support from the NPs surface after repeated reaction cycles which eventually results in agglomeration of the particles and decrease in catalytic activity. Alternatively, nanoparticles have been immobilized or grafted onto solid supports to improve their stabilization and recycling ability [8]. "Inert" porous solid materials such as zeolites, charcoals, metal-organic frameworks, layered-double hydroxides etc. have been used for immobilization of active metal catalysts. Being inert and hard, these materials are highly stable even under harsh reaction conditions and can be easily recovered from the reaction mixture for further applications. Porous materials also allow control over the nanoparticle growth in the porous matrix, prevents agglomeration and enhances active site exposure [11]. The host support materials stabilizes the nanoparticles, however they have no role in manipulating the activity of the nanoparticles for catalysis. In the last few years, there has been tremendous focus on the development of "active" supports, which along with stabilizing the nanoparticles also contribute towards overall catalytic activity in synergy with the nanoparticles [12, 13]. For example, the charge state of the Au nanoparticles is known to influence their reactivity, in the case of the negatively charged Au nanoparticles, an extra electron from the gold readily transfers to the anti-bonding $2\pi *$ orbital of the adsorbed O_2, which weakens the O—O bond and activates oxygen molecule for further catalytic reaction. On the other hand, the positive charge accumulated on the gold can promote adsorption of some reactants, such as CO and hydrocarbons. An active support can transfer charges to/from the active catalytic surface, hence influencing the activity of the reaction. For example, Au nanoparticles anchored on rutile TiO_2 (110) surface shows high activity for the oxidation of CO to CO_2. Along with providing significant exposed catalytic active sites for the reaction, TiO_2 also involves in charge transfer process with the Au NPs making the NPs surface highly negative for dioxygen activation leading to oxidation of CO to CO_2 [14]. Recently, research works involving GO as a support for immobilizing active metal nanoparticles have gained attention. GO, not only provides a large surface area with high exposure of active catalysts, but also can influence the catalytic activity [15]. Possible surface to metal electron transfer from GO to nanoparticles activating dioxygen molecule over NPs surface for several oxidation reactions has been reported [15]. Hence, choice of a suitable support for NPs stabilization with possible cooperativity might play an important role in controlling the reaction yield and selectivity of products.

2. Carbocatalysis

Carbon is one of the most abundant elements on earth and is central to life. Hence, catalytic application of carbon is very attractive and both organic and inorganic carbons play a key role in catalysis. A huge amount of organic compounds act as highly efficient homogeneous catalysts, forming a dedicated branch of chemistry "organocatalysis." Carbon is often the main constituent of the organic ligands surrounding the metallic center in organometallics. In enzymatic catalysis it constitutes the backbone of the active species. In heterogeneous catalysis, carbon materials act as unique catalyst supports by anchoring different active species through its active site and can also act as catalysts by themselves. The physical and chemical properties of carbon materials, such as their tunable porosity and surface chemistry, make them suitable for application in many catalytic processes.

Among the carbon catalysts developed, activated carbon (AC) and carbon black (CB) are the most commonly used carbon supports. The typically large surface area and high porosity of activated carbon catalysts favor the dispersion of the active phase over the support and increase its resistance to sintering at high metal loadings. The pore size distribution can be tuned to suit the requirements of active supports and substrates. The activated carbon shows several advantages owing to their several outstanding properties, such as low cost, resistance to acids and bases, high stability even at elevated temperature, high surface area (>1000 m^2/g)

and easy removal etc. Moreover, metal salts can be reduced to active metallic forms in these mesoporous materials, making them highly competent as metal supports.

The study of chemical reactions using carbon materials are termed as carbocatalysis. The catalysts are prepared and used in the powder form and hence they are heterogeneous. Carbocatalysis has been known for decades since the first discovery of catalytic activities of carbon materials [16] when Rideal and Wright [17] showed the charcoal catalyzed oxidation of oxalic acid, which was one of the ground breaking discovery of carbocatalysis. Moreover, 45 years earlier also, carbon materials were shown to be effective for the conversion of halogenated hydrocarbon [16].

With the development of fullerenes, the research activities for the growth of nanocarbon materials have gained momentum. Several polytypes of carbon which include fullerenes, nanotubes, graphene, nanodiamonds and amorphous porous carbon and their derivatives represent a rich class of solid carbonaceous materials with environmental acceptability and reusability and all are found to be catalytically active in certain reactions. However, most of these carbon materials are highly hydrophobic without any functional groups on their surface.

Fullerene black is an efficient catalyst for dehydrogenation, cracking, methylation, and demethylation reactions. C_{60} and C_{70} were found to be suitable catalysts for the reduction of nitrobenzene, using hydrogen gas under UV light [18]. Further, several organometallic compounds involving fullerene as a ligand have been developed that showed efficient catalytic activity for several organic transformations.

The 1D and 2D carbon materials such as carbon nanotubes and graphene offered high surface area and continuous efforts are focused on surface functionalization of these materials, both through covalent and non-covalent approach. Oxidation in presence of strong acids and oxidants could introduce oxyfunctionalized groups on the surfaces of these carbon materials, making them hydrophilic and suitable for anchoring several active catalytic groups on their surfaces. The work on the oxidative dehydrogenation reaction by Mestl et al. [19] and Zhang et al. [20] opened a new window in carbocatalysis. Carbon nanotubes, in its oxygenated forms, showed efficient catalysis for oxidative dehydrogenations e.g. conversion of n-butane to 1-butene [20]. In the catalytic hydrogenation of ethylbenzene to styrene, a process of high industrial relevance, CNTs perform better than activated carbon and graphite as catalysts. It was reasoned that the reactant molecules were first adsorbed on the CNT surface via π-interactions next to basic oxygen moieties, which facilitated dehydrogenation with concomitant formation of surface hydroxyl groups [19]. Taking advantage of surface modification techniques, various nanoparticle as well as molecular catalysts could be anchored on carbon nanotubes [21] (**Figure 1**).

2.1. Graphene oxide as a carbocatalyst

Graphene and other two-dimensional sp^2-hybridized carbon scaffolds are expected to have large impacts in the area of catalysis, mainly because of their unique electronic properties

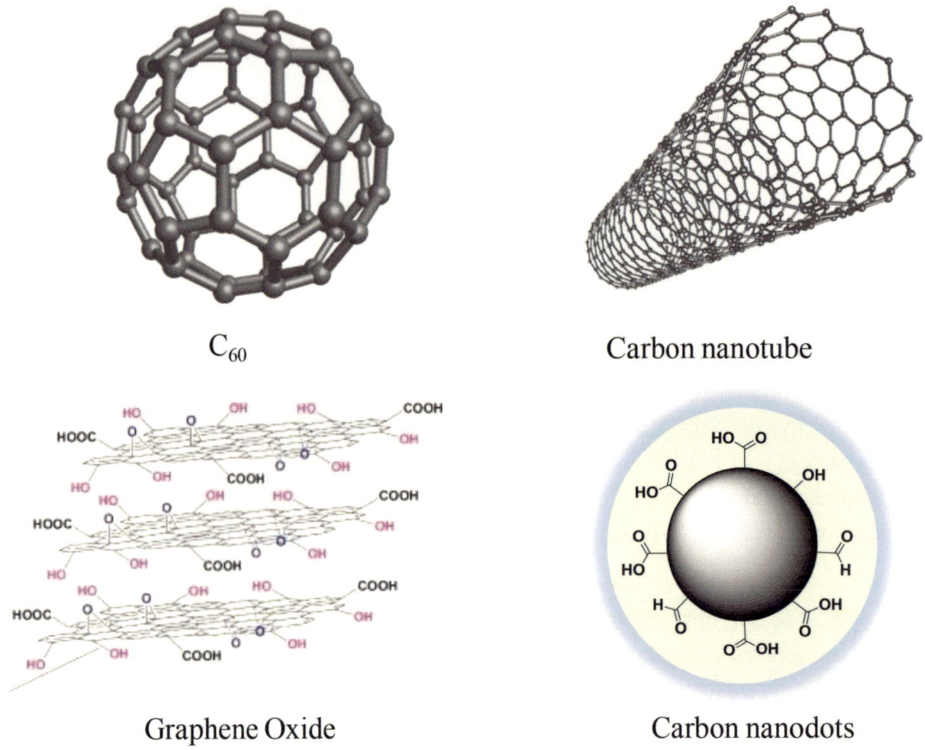

Figure 1. Various forms of carbon nanomaterials.

and high surface area in comparison to other carbon materials [22]. Although graphene was known to exist within graphite materials, it was assumed to be thermodynamically unstable in distinct 2D structures at finite temperatures. Geim et al. (2004) [23] mechanically exfoliated single sheets from the π-stack layers in graphite for the first time. The unique electron transfer properties of graphene, such as a half-integer quantum Hall effect, the massless Dirac fermion behavior of its charge carriers, and quantum capacitance, have been extensively studied making them one of the most important materials in optoelectronics utility. The use of graphene-based nanomaterials as catalyst support was hampered by the high price associated with the laborious synthesis and processing (e.g., sublimation of silicon from silicon carbide wafers, chemical vapor deposition, oxidation/reduction protocols etc. However, the process for liquid phase exfoliation through oxidation of graphite in presence of strong oxidizing agents generating the graphene analogue with oxygenated functionalities on their surface (popularly known as Hummer's method) has brought tremendous excitement in the nanocatalyst research community. These materials termed as "graphene oxide" can be obtained in sufficient quantities from commercially available graphite through reliable, now well-established preparation procedures. Further potential chemical modifications of the graphene surface introduces different newer catalytically active site important for specific catalytic reactions (**Figure 2**) [23].

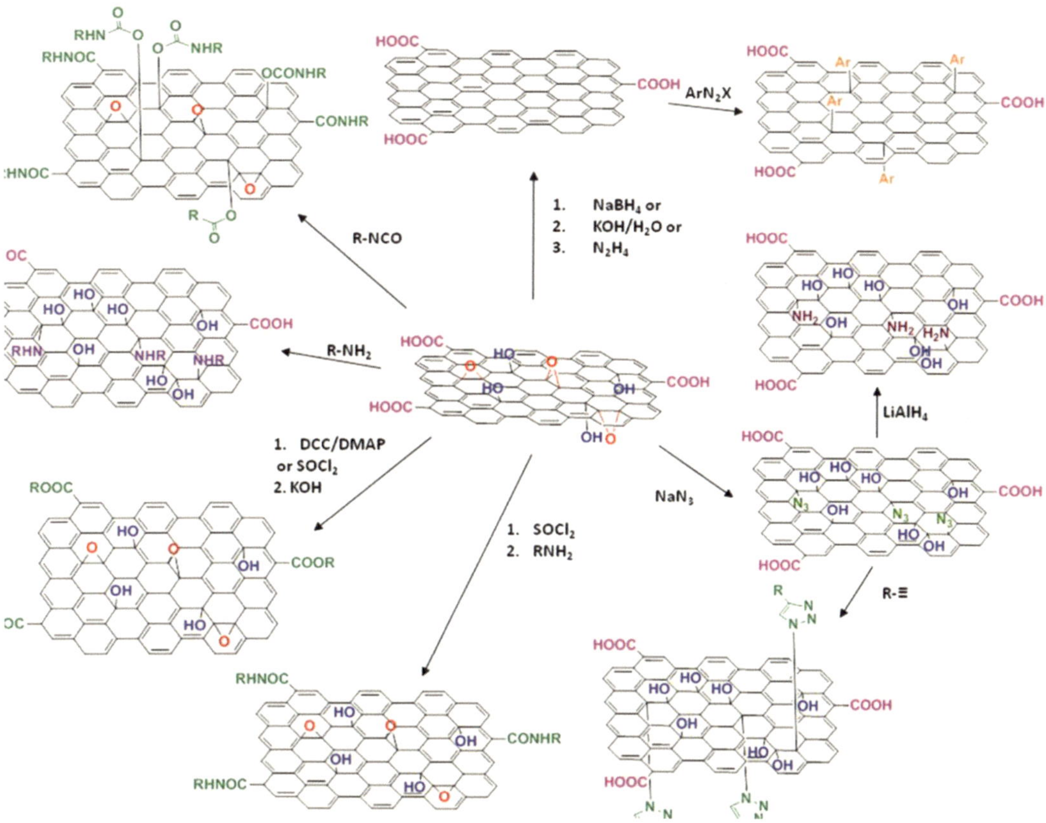

Figure 2. Possibilities of covalent functionalization of GO (reprinted with permission from Ref. [23]. American Chemical Society).

Oxidation of graphite leads to the decoration of the graphene surface with oxygen functionalities that increases the inter-layer separation, thus helping in exfoliation into single or a few layer two-dimensional surfaces. During this process, several sites are induced those are important from catalytic or surface modification point of view. The extensive π-conjugated domains provide interactions between aromatic compounds with the graphene surfaces and greatly facilitate the adsorption/activation of aromatic compounds on graphene based carbon [23]. There are at least five different oxygen functional groups decorated over the graphene surface. These include carboxyl (—COOH), hydroxyl (—OH), carbonyl (—C=O), epoxy (—C—O—C— and ketone (—C=O) groups. These oxygenated groups provide four different categories of catalytic activity to the carbon material: (1) their acidic properties promote acid-catalyzed reactions; (2) their intermediate form reacts with oxidants to catalyze oxidation reaction; (3) their nucleophilic nature promote coupling reactions; and (4) their perfect π-conjugated structure with significant defects/holes can also promote several catalytic reactions. Further, reduction of graphene oxides can be performed using common reducing agents such as hydrogen, metal ion borohydride and hydrazine. Moreover, the graphene oxide surfaces can be reduced by heating at elevated temperature. Various heteroatoms such as N, B, P, Se, S, F, and Cl [24] can be incorporated into the lattice of graphene sheets. Several organic reactions can also incorporate acidic functional groups such as $-SO_3H$ groups onto graphene sheets [25].

GO and their chemically converted forms have shown broad spectrum of catalytic activity ranging from oxidation reactions and thermal decomposition reactions. Bielawski et al., first demonstrated catalytic activity of graphene oxide for liquid phase organic transformations [26]. Since then, a variety of organic transformations have been explored taking advantage of the functional groups present on the graphitic surface. **Table 1** summarizes a variety of reported reactions catalyzed by GO and chemically converted GO.

Further, the two-dimensional surface of graphene based materials can be used to anchor other active catalysts as well as biocatalysts. For example, the catalytic activity of several enzymes including cytochromes, peroxidases, myoglobins, and hemoglobins supported on graphene

Catalyst	Reactions	Active sites	References
	Oxidation reactions (promoted by molecular O_2)		
(BN)HolG	Aerobic oxidation of amines	Doped N and B are active sites	[28]
GO	Aerobic oxidation of benzylic alcohols	Oxygen functional groups	[29]
	Oxidation reactions (promoted by other oxidants)		
rGO	Oxidation of pollutants (H_2O_2)	Electron rich oxygen groups	[30]
	Reduction reactions		
B, N or O doped G ribbon edges	H_2 dissociative adsorption chemisorption on G ribbon edge	Beneficial B doping	[31]
	Acid reactions		
rGO-SO_3H	Dehydration of xylose to furfural	$-SO_3H$ are active sites	[32]
	Base reactions		
rGO-NEt_3	Hydrolysis of ethyl acetate	Amino groups are active sites	[33]
rGO-PAMAM	Knoevenagel condensation	Basic sites of the catalyst	[34]
	Thermal decomposition reaction		
rGO	Thermal dehydrogenation and rehydrogenation of $LiBH_4$		[35]

Table 1. Catalytic reactions by GO and chemically converted GO. (B, N)-doped holy G [(BN)HolG], reduced graphene oxide (rGO), triethylamine modified rGO (rGO-NEt_3), rGO functionalized with $-SO_3H$ (rGO-SO_3H) and poly(amidoamine)-modified rGO (rGO-PAMAM).

surface enhances several folds compared to the unsupported enzymes towards the oxidation reaction of pyrogallol [27]. The importance of oxygen functional groups on GO surface has been exploited towards several C—H activation and C—C coupling reactions. The carbocatalytic activity of graphene oxide has successfully been exploited by Ma et al. for the C—H bond arylation of benzene enabling biaryl construction. The oxygen functional groups in these graphene oxide sheets and presence of KO^{t}Bu were demonstrated to be essential for the observed catalytic activity. Several model reactions and DFT calculations confirmed that the negatively charged oxygen atoms promote the overall transformation by stabilizing and activating K^+ ions, which in turns facilitates the activation of the C—I bond. The π basal plane also greatly facilitates the overall reaction as the aromatic coupling partners are easily adsorbed on the 2d surface [36]. Transition-metal-catalyzed alkylation reactions of arenes have turn out to be a central transformation in organic synthesis. Szostak et al. developed the first general strategy for alkylation of arenes with styrenes and alcohols catalyzed by carbon-based materials, exploiting the unique surface property of graphene oxide to produce valuable diarylalkanes with excellent yields and regioselectivity. Remarkably, this protocol represents the first general application of graphene oxide to promote direct C—C bond formation utilizing oxygenated functional groups present on GO surface [37]. Recently, our research group have demonstrated the dual carbocatalytic activity of graphene oxide for the C—N coupling reaction towards the formation of α-ketoamides through a cross-dehydrogenative coupling pathway [38]. The presence of polar functional groups (e.g., carboxyl, hydroxyl, ketonic, and epoxides) on graphene oxide surface induce acidic as well as oxidizing properties to the material. This dual catalytic property of the material is explored towards the generation of α-ketoamides where surface acidity favors the initial formation of hemiaminal intermediate followed by oxidation leading to the desired final product. Several control experiments as well as thermal treatment showed that it is the oxygen functional groups, especially carboxylic acid group that is only responsible for the observed catalytic activity. The protocol could also be extended towards the synthesis of biologically important α-ketoamides. On the other hand graphene surface can also be used as support for immobilization of several metal/metal oxide nanoparticles and used for several electrocatalysis, photocatalysis and organic transformations [39, 40]. For example, Pd nanoparticle immobilized on graphene oxide gave remarkable turnover frequencies (TOF > 39,000 h^{-1}) in Suzuki-Miyaura cross-coupling reactions. Microwave assisted reduction of well-dispersed GO and palladium salt to form Pd/rGO [41] demonstrated outstanding catalytic activity for the Suzuki-Miyaura coupling reaction (TOF up to 108,000 h^{-1}) under ligand-free conditions, which was attributed to the high concentration of well dispersed Pd-NPs.

3. Current and future prospect

The carbon based nanomaterials have already demonstrated their enormous potential either as catalysts or heterogeneous catalyst supports. Graphene oxide with oxygenated functional groups on their surface could act as active sites for various acid catalyzed and oxidative catalytic reactions. Recent advancement of these graphene based materials shows that the modification of graphene surface by different methods leads to generation of holes which acts as traps for reactive oxygen species for many challenging organic reactions [42]. Hence tremendous possibilities of these carbonaceous nanomaterials remain to explore in various fields including chemical synthesis, energy storage, fuel-cells, environmental remediation and organism degradation. Carbon nanodots are the recent inclusion to the nanocarbon family. The excellent photoluminescence properties of carbon nanodots have directed their application in different fields including sensing, optoelectronics, bio imaging, nanomedicine, etc. Although they are widely explored in sensing as well as bio-medical application, their inherent photocatalytic capability towards organic synthesis has not been explored much. So, the development of carbon nanodots towards organic synthesis may result in an important alternative to the traditional transition metal based catalysts. There are still huge scope towards (i) high performance carbon catalyst specific for desired products, (ii) development of chiral carbon nanomaterials for enantioselective synthesis, (iii) affordable methods for large scale synthesis, industrial scalability and economic viability, (iv) detailed elucidation of catalytic mechanism that can bring further improvements in catalytic activity and (v) stability of the catalyst to maintain excellent catalytic activity during recycling. Overall, development of carbon related catalysts with broader applications is imminent towards green and sustainable chemistry.

4. Conclusion

In this chapter, we have reviewed the inherent carbocatalytic activity of graphene oxide towards different organic transformations. The π-π^* network as well as oxygenated functional groups present over the GO surface both contributed towards the enhanced catalytic activity. The presence of π-π^* network on graphene surface helped the substrate molecules getting adsorbed over the catalyst surface and the oxygenated functional groups catalyzed the reaction. The oxygenated functional groups acted as the active site for several acid catalyzed as well as oxidation reactions. Further, the presence of various oxygenated functionalities on the surface of graphene oxide could be used for anchoring other active catalysts such as metal and metal oxide nanoparticles. The reducing capability of these nanocarbons could also be envisaged for reduction of metal ions to nanoparticles.

Author details

Biju Majumdar, Daisy Sarma and Tridib K. Sarma*

*Address all correspondence to: tridib@iiti.ac.in

Discipline of Chemistry, Indian Institute of Technology Indore, Indore, Madhya Pradesh, India

References

[1] Chorkendorff I, Niemantsverdriet JW. Concepts of Modern Catalysis and Kinetics. 2nd ed. KGaA, Weinheim: Wiley-VCH Verlag GmbH & Co; 2007

[2] Gladysz JA. Introduction: Recoverable catalysts and reagents-perspective and prospective. Chemical Reviews. 2002;**102**:3215-3216. DOI: 10.1021/cr020068s

[3] Schmid G. Large clusters and colloids. Metals in the embryonic state. Chemical Reviews. 1992;**92**:1709-1727. DOI: 10.1021/cr00016a002

[4] Lewis LN. Chemical catalysis by colloids and clusters. Chemical Reviews. 1993;**93**:2693-2730. DOI: 10.1021/cr00024a006

[5] Daniel MC, Astruc D. Gold nanoparticles: Assembly, supramolecular chemistry, quantum-size-related properties, and applications toward biology, catalysis, and nanotechnology. Chemical Reviews. 2004;**104**:293-346. DOI: 10.1021/cr030698

[6] Dahl JA, Maddux BLS, Hutchinson JE. Toward greener nanosynthesis. Chemical Reviews. 2007;**107**:2228-2269. DOI: 10.1021/cr050943k

[7] Rao CNR, Muller A, Cheetham AK. The Chemistry of Nanomaterials: Synthesis and Applications. Vol. 1. Weinheim: Wiley-VCH; 2004

[8] Martino A, Yamanaka SA, Kawola JS, Loy DA. Encapsulation of gold nanoclusters in silica materials via an inverse micelle/sol-gel synthesis. Chemistry of Materials. 1997;**9**:423-429. DOI: 10.1021/cm9604625

[9] Lu A-H, Salabas EL, Schüth F. Magnetic nanoparticles: Synthesis, protection, functionalization, and application. Angewandte Chemie, International Edition. 2007;**46**:1222-1244. DOI: 10.1002/anie.200602866

[10] Somorjai GA, Park JY. Molecular factors of catalytic selectivity. Angewandte Chemie, International Edition. 2008;**47**:9212-9228. DOI: 10.1002/anie.200803181

[11] Mirkin CA. The beginning of a small revolution. Small. 2005;**1**:14-16. DOI: 10.1002/smll.200400092

[12] Wang Y-J, Wilkinson DP, Zhang J. Noncarbon support materials for polymer electrolyte membrane fuel cell electrocatalysts. Chemical Reviews. 2011;**111**:7625-7651. DOI: 10.1021/cr100060r

[13] Vernoux P, Lizarraga L, Tsampas MN, et al. Ionically conducting ceramics as active catalyst supports. Chemical Reviews. 2013;**113**:8192-8260. DOI: 10.1021/cr4000336

[14] Wang Y-G, Yoon Y, Glezakou V-A, Li J, Rousseau R. The role of reducible oxide-metal cluster charge transfer in catalytic processes: New insights on the catalytic mechanism of CO oxidation on Au/TiO_2 from ab initio molecular dynamics. Journal of the American Chemical Society. 2013;**135**:10673-10683. DOI: 10.1021/ja402063v

[15] Majumdar B, Bhattacharya T, Sarma TK. Gold nanoparticle-polydopamine-reduced graphene oxide ternary nanocomposite as an efficient catalyst for selective oxidation of benzylic $C(sp^3)$—H bonds under mild conditions. ChemCatChem. 2016;**8**:1825-1835. DOI: 10.1002/cctc.201600136

[16] Rodríguez-Reinoso F. The role of carbon materials in heterogeneous catalysis. Carbon. 1998;**36**:159-175. DOI: 10.1016/S0008-6223(97)00173-5

[17] Rideal EK, Wright WM. CLXXXIV-low temperature oxidation at charcoal surfaces. Part I. The behavior of charcoal in the absence of promoters. Journal of the Chemical Society, Transactions. 1925;**127**:1347-1357. DOI: 10.1039/CT9252701347

[18] Li B, Xu Z. A non-metal catalyst for molecular hydrogen activation with comparable catalytic hydrogenation capability to noble metal catalyst. Journal of the American Chemical Society. 2009;**2009**(131):16380-16382. DOI: 10.1021/ja9061097

[19] Mestl G, Maksimova NI, Keller N, Roddatis VV, Schlögl R. Carbon nanofilaments in heterogeneous catalysis: An industrial application for new carbon materials? Angewandte Chemie, International Edition. 2001;**40**:2066-2068. DOI: 10.1002/1521-3773(20010601)40:11<2066::AID-ANIE2066>3.0.CO;2-I

[20] Zhang J, Liu X, Blume R, Zhang A, Schlögl R, Su DS. Surface-modified carbon nanotubes catalyze oxidative dehydrogenation of n butane. Science. 2008;**322**:73-77. DOI: 10.1126/science.1161916

[21] Basu P, Prakash P, Gravel E, Shah N, Bera K, Doris E, Namboothiri INN. Carbon nanotube-ruthenium hybrids for the partial reduction of 2-nitrochalcones: Easy access to quinoline N-oxides. ChemCatChem. 2016;**8**:1298-1302. DOI: 10.1002/cctc.201600042

[22] Pyun J. Graphene oxide as catalyst: Application of carbon materials beyond nanotechnology. Angewandte Chemie, International Edition. 2011;**50**:46-48. DOI: 10.1002/anie.201003897

[23] Novoseleov KS, Geim AK, Morozov SV, Jiang D, Zhang Y, Dubunos SV, Grigorieva IV, Firsov AA. Electric field effect in atomically thin carbon films. Science. 2004;**306**:666-669. DOI: 10.1126/science.1102896

[24] Cheng Y, Fan Y, Pei Y, Qiao M. Graphene-supported metal/metal oxide nanohybrids: Synthesis and applications in heterogeneous catalysis. Catalysis Science & Technology. 2015;**5**:3903-3916. DOI: 10.1039/C5CY00630A

[25] Ji J, Zhang G, Chen H, Wang S, Zhang G, Zhang F, Fan X. Sulfonated graphene as water-tolerant solid acid catalyst. Chemical Science. 2011;**2**:484-487. DOI: 10.1039/C0SC00484G

[26] Dreyer DR, Jia HP, Bielawski CW. Graphene oxide: A convenient carbocatalyst for facilitating oxidation and hydration reactions. Angewandte Chemie, International Edition. 2010;**49**:6813-6816. DOI: 10.1002/anie.201002160

[27] Xue T, Jiang S, Qu Y, Su Q, Cheng R, Dubin S, Duan X. Graphene-supported hemin as a highly active biomimetic oxidation catalyst. Angewandte Chemie, International Edition. 2012;**51**:3822-3825. DOI: 10.1002/anie.201108400

[28] Li X-H, Antonietti M. Polycondensation of boron- and nitrogen-Co doped holey graphene monoliths from molecules: Carbocatalysts for selective oxidation. Angewandte Chemie, International Edition. 2013;**52**:4572-4576. DOI: 10.1002/anie.201209320

[29] Boukhvalov DW, Dreyer DR, Bielawski CW, Son Y-W. A computational investigation of the catalytic properties of graphene oxide: Exploring mechanisms by using DFT methods. ChemCatChem. 2012;**4**:1844-1849. DOI: 10.1002/cctc.201200210

[30] Peng W, Liu S, Sun H, Yao Y, Zhi L, Wang S. Synthesis of porous reduced graphene oxide as metal-free carbon for adsorption and catalytic oxidation of organics in water. Journal of Materials Chemistry A. 2013;**1**:5854-5859. DOI: 10.1039/C3TA10592J

[31] Liao T, Sun C, Sun Z, Du A, Smith S. Chemically modified ribbon edge stimulated H_2 dissociation: A first-principles computational study. Physical Chemistry Chemical Physics. 2013;**15**:8054-8057. DOI: 10.1039/c3cp50654a

[32] Lam E, Chong JH, Majid E, Liu Y, Hrapovic S, Leung ACW, Luong JHT. Carbocatalytic dehydration of xylose to furfural in water. Carbon. 2012;**50**:1033-1043. DOI: 10.1016/j.carbon.2011.10.007

[33] Yuan C, Chen W, Yan L. Amino-grafted graphene as a stable and metal-free solid basic catalyst. Journal of Materials Chemistry. 2012;**22**:7456-7460. DOI: 10.1039/C2JM30442B

[34] Wu T, Wang X, Qiu H, Gao J, Wang W, Liu Y. Graphene oxide reduced and modified by soft nanoparticles and its catalysis of the Knoevenagel condensation. Journal of Materials Chemistry. 2012;**22**:4772-4779. DOI: 10.1039/C2JM15311D

[35] Xu J, Meng R, Cao J, Gu X, Qi Z, Wang W, Chen Z. Enhanced dehydrogenation and rehydrogenation properties of $LiBH_4$ catalyzed by graphene. International Journal of Hydrogen Energy. 2013;**38**:2796-2803. DOI: 10.1016/j.ijhydene.2012.12.046

[36] Gao Y, Tang P, Zhou H, Zhang W, Yang H, Yan N, Ma D. Graphene oxide catalyzed C—H bond activation: The importance of oxygen functional groups for biaryl construction. Angewandte Chemie, International Edition. 2016;**55**:3124-3128. DOI: 10.1002/anie.201510081

[37] Hu F, Patel M, Luo F, Flach C, Mendelsohn R, Garfunkel E, Szostak M. Graphene-catalyzed direct Friedel-Crafts alkylation reactions: Mechanism, selectivity, and synthetic utility. Journal of the American Chemical Society. 2015;**137**:14473-14480. DOI: 10.1021/jacs.5b09636

[38] Majumdar B, Sarma D, Bhattacharya T, Sarma TK. Graphene oxide as metal-free catalyst in oxidative dehydrogenative C—N coupling leading to α-ketoamides: Importance of dual catalytic activity. ACS Sustainable Chemistry & Engineering. 2017;**5**:9286-9294. DOI: 10.1021/acssuschemeng.7b02267

[39] Devadoss A, Sudhagar P, Das S, Lee SY, Terashima C, Nakata K, Paik U. Synergistic metal–metal oxide nanoparticles supported electrocatalytic graphene for improved photo electrochemical glucose oxidation. ACS Applied Materials & Interfaces. 2014;**6**:4864-4871. DOI: 10.1021/am4058925

[40] Guo XN, Jiao ZF, Jin GQ, Guo XY. Photocatalytic Fischer-Tropsch synthesis on graphene-supported worm-like ruthenium nanostructures. ACS Catalysis. 2015;**5**:3836-3840. DOI: 10.1021/acscatal.5b00697

[41] Moussa S, Siamaki AR, Gupton BF, El-Shall MS. Pd-partially reduced graphene oxide catalysts (Pd/PRGO): Laser synthesis of Pd nanoparticles supported on PRGO nanosheets for carbon-carbon cross coupling reactions. ACS Catalysis. 2012;**2**:145-154. DOI: 10.1021/cs200497e

[42] Su C, Acik M, Takai K, Lu J, Hao S-J, Zheng Y, Wu P, Bao Q, Enoki T, Chabal YJ, Loh KP. Probing the catalytic activity of porous graphene oxide and the origin of this behaviour. Nature Communications. 2012;**3**:1298. DOI: 10.1038/ncomms2315

Chapter 4

Graphene Oxide/Reduced Graphene Oxide Aerogels

Gudkov Maksim Vladimirovich and
Valery Pavlovich Melnikov

Additional information is available at the end of the chapter

http://dx.doi.org/10.5772/intechopen.78987

Abstract

In this chapter, we will discuss aerogels based on graphene oxide/reduced graphene oxide—promising composite materials, based on 2D carbon nanoparticles, with a certain architecture that prevents aggregation of graphene layers with a highly developed surface that have a high potential technological realization as materials for supercapacitors, sensors, batteries, and actuators. This chapter describes the existing methods for producing composite aerogels based on graphene oxide/reduced graphene oxide, the structural features of these materials, the most relevant studies in the areas of surface modification, architectural control, improvement of mechanical properties, and the most interesting applications. It has been shown that such materials have relatively high specific surface values and a high degree of exfoliation of graphene layers, but an urgent need is to improve them, which is due to the fragility of graphene aerogels and composite materials based on them, as well as the need to modify the surface to control porosity.

Keywords: graphene oxide aerogels, reduced graphene oxide aerogels, self-gelation, cross linking, hydrothermal synthesis, highly porous materials, carbon based materials

1. Introduction

At present, one of the most urgent topics in science is the creation of aerogels based on reduced graphene oxide. Their high porosity, characterized by low material density and high specific surface area, as well as the ability to conduct electric current attracts the attention of various groups of scientists. The work on the creation of such materials has been conducted for the past 20 years. A number of materials developed during this time and approaches to their production form a serious prospect for the use of these materials as supercapacitors, gas sensors, electric batteries, and actuators. The increased interest in graphene oxide as a precursor of such materials is due to its ability to form stable colloidal dispersions in polar solvents (such as water) and also to transform into a reduced electrically conductive graphene-like form under chemical or thermal treatment. In addition, the ability of such materials to multi-cyclic deformations, flexibility, and elasticity attracts special interest of researchers. The rapid growth of publication activity in this theme in recent years is partly due to the need for the development of portable electronic devices, in particular, energy storage devices that are included in the everyday life of each person. In the development of aerogels based on reduced graphene oxide, the authors are striving to increase the electrical conductivity, the specific surface area, the mechanical strength, elasticity, and durability of these materials. The part of the science effort was aimed at studying the methods of obtaining aerogels from reduced graphene oxide. It has been shown that the main and most effective methods for the formation of superporous ultralight structures are self-gelation of graphene oxide containing systems, hydrothermal-assisted formation of aerogels, cross-linking of a structure, followed by freeze drying or supercritical drying. Each of the described methods has its advantages and disadvantages, which are effectively being used to achieve the target parameters of synthesized materials. To improve the obtained structures and give them new properties, the surface of graphene-like sheets is being modified in various ways: various polymeric materials are being introduced into the structure, the surface is being decorated with metal particles, the materials are being doped with nitrogen, etc. The application of these approaches made it possible to significantly improve the mechanical, electrophysical, catalytic, and sorption properties of aerogels and allow obtaining materials with controlled architecture and unique morphology of the surface. The latest achievements in this area will be discussed in this chapter.

2. Synthesis of reduced graphene oxide aerogels

Unique structures and attractive properties of reduced graphene oxide (rGO)-based 3D materials were the reasons to establish a number of approaches for fabrication of those materials with controlled regular structure. The main methods of organization of such structures are self-gelation of graphene oxide (GO) containing systems, hydrothermal-assisted formation of aerogels, and cross-linking of a structure. Almost all these methods need a support of freeze drying or supercritical drying due to ability to prevent stacking of graphene oxide/reduced graphene oxide sheets during drying process. The absence of mobility of the GO/RGO layers and the fixation of the dispersion structure make it possible to obtain materials with a high porosity, a high specific surface area, and a high degree of exfoliation of the graphene-like sheets in the final material.

As a rule, freeze drying of the samples leads to the formation of a large number of macropores. This is due to the fact that during the freezing of the sample, ice crystals are formed, which displace part of the material from its volume, which leads to local structural irregularities and the formation of macropores in place of the crystals after they are removed from the matrix during the drying process. It is clear that the size of the ice crystals is highly dependent on the freezing speed of the material, but even the use of liquid nitrogen for freezing does not avoid macroporosity due to the low thermal conductivity of nitrogen. Supercritical drying leads to the formation of a uniform microporous structure. Both approaches lead to the formation of ultralight porous structures with highly developed morphology, but each of them has its advantages. For example, the presence of macropores provides a large number of diffusion paths for various molecules and ions, which is an undoubted advantage for a number of applications.

Various organic and inorganic reducing agents are used. Zhang et al. used L-ascorbic acid to obtain a mechanically strong and electrically conductive aerogel [1]. The authors showed that, in contrast to other reducing agents ($NaBH_4$, $LiAlH_4$, hydrazine), the reduction with L-ascorbic acid does not lead to the formation of reaction by-products that significantly affect the uniformity of the structure of the synthesized aerogel. $NaHSO_3$, vitamin C, Na_2S, ammonia boron trifluoride, sodium ascorbate, and hydroquinone have also been used by various investigators to form aerogels based on reduced graphene oxide [2, 3]. Aerogels based on reduced graphene oxide have also been obtained using a new inexpensive, environmentally friendly reducing medium that combines oxalic acid and sodium iodide [4]. The materials showed a low density, high porosity structure, and electrical conductivity. Yang et al. have developed a light, environmentally friendly, mild method of forming aerogel by thermal evaporation of a suspension of graphene oxide in the presence of $NaHCO_3$ [5]. This approach is based on in-situ reduction to form an rGO aerogel. The aerogel based on the reduced graphene oxide was also synthesized using the hypophosphorous acid reduction process and I_2 [6]. He showed a large surface area of 830 m^2/g. Typically, the reduction with chemical agents is carried out in a liquid medium; however, in work [7], an interesting approach was first proposed for the reduction of the aerogel of graphite oxide with hydrazine vapor at room temperature after drying. The authors showed that the reduction has explosive character, and the resulting material practically does not differ from the material obtained by thermal shock reduction at 600°C (**Figure 1**). Also, various organic amines are used as reducing agents.

2.1. Gelation of graphene oxide followed by drying

The simplest and most effective way to produce three-dimensional materials from graphene-like sheets is to obtain a hydrogel from graphene oxide, followed by removal of the solvent by drying. It is known that a stable dispersion of graphene oxide is formed at a certain concentration [8] due to the optimal ratio of the forces of electrostatic repulsion due to the huge number of oxygen-containing functional groups on the basal plane of the particles and the van der Waals interaction of monolayers. Gelation of suspensions occurs when the balance between these forces is violated and leads to the stratification of the particles of graphite oxide on the monolayers of graphene oxide (**Figure 2**).

The process of gel formation depends on many factors: van der Waals interaction, n-n stacking, the formation of hydrogen bonds, and electrostatic interaction [9]. The most important factor is also the concentration of suspensions, but the data on the optimal concentration are quite different. It has been shown that the optimal concentrations of gelation when exposed to ultrasound in a suspension of GO are 30 mg/ml and 0.075–0.125 mg/ml in different cases [10], but the aerogel obtained in the second case had a weak mechanical strength. In the work of Bai et al. [11], the optimal concentration was 4 mg/ml. Such a strong difference in the results, in our opinion, is due to several factors. The first of these is the method for obtaining

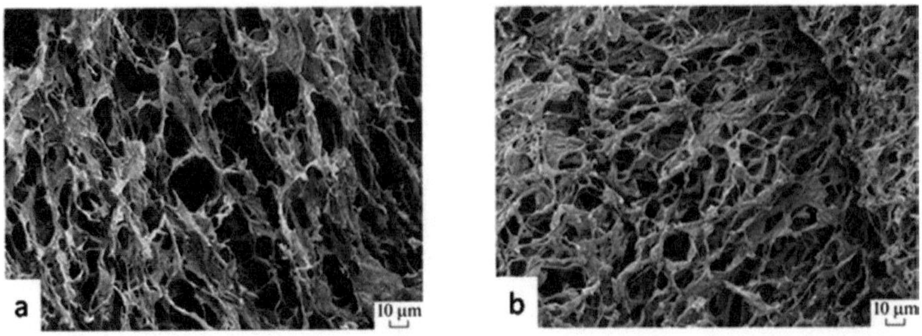

Explosion initiator	T, °C	CO_2	H_2O	CO	N_2	Total
		mmol/g				
Hydrazine	23–25	2.5	2.3	4.6	1.3	10.7
Thermal shock	600	3.4	4.3	3.6	–	11.3

c

Explosion initiator	C	O	H	N	C/O
	at %				
Hydrazine	82.9	12.7	3.5	0.9	6.5
Thermal shock	83.0	12.0	5.0	–	6.9

d

Figure 1. Micrographs of materials obtained by explosive reduction of graphite oxide aerogel by (a) thermal shock and (b) hydrazine vapor. (c) Gaseous products of explosive reduction of graphite oxide aerogel by thermal shock and by hydrazine vapor. (d) Comparison of the elemental compositions of materials obtained by explosive reduction of graphite oxide aerogel by thermal shock and by hydrazine vapor [7].

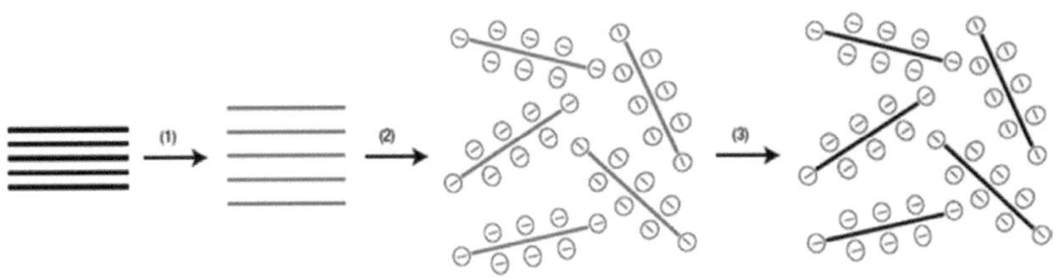

Figure 2. Scheme showing the chemical route to the synthesis of aqueous graphene dispersions. 1, Oxidation of graphite (black blocks) to graphite oxide (lighter colored blocks) with greater interlayer distance. 2, Exfoliation of graphite oxide in water by sonication to obtain GO colloids that are stabilized by electrostatic repulsion. 3, Controlled conversion of GO colloids to conducting graphene colloids through deoxygenation by hydrazine reduction [8].

a dispersion of graphene oxide. In the case where the critical gelling concentration is high, the dispersion is typically prepared by redispersing the sample of graphite oxide (in the form of a film or powder) air dried at an elevated temperature (typically T = 60°C). As practice shows, this method leads to the staking of monolayers and their redispergation is greatly hampered, which leads to a significant decrease in the ability to form a spatial grid and to gel. On the contrary, if the dispersion is prepared bypassing the drying stage of the product under such conditions (for example, by purifying the product from acids by dialysis), the self-gelation of dispersion occurs at substantially lower concentrations, even in the absence of ultrasonic treatment. The second factor is the size of the basal plane of the particles of graphene oxide, which is rarely taken into account in the processes of self-gelation of GO dispersions. This is due to the fact that it is rather difficult to obtain particles of graphene oxide with a given basal plane size. It is clear that the interaction of particles of different sizes should be completely different, but there are no dependencies of this kind in the scientific literature. The third factor is, at times, quite a noticeable difference, including in the chemical composition, of the graphene oxide itself (**Figure 3**), obtained from different types of graphite [12].

2.2. Hydrothermal-assisted formation of reduced graphene oxide aerogel

Hydrothermal synthesis is a method of obtaining various chemical compounds and materials using physicochemical processes in closed systems that occur in aqueous solutions at temperatures above 100°C and pressures above 1 atm. In graphene systems, this process is realized at temperatures close to the effective reduction temperature of graphene oxide (200°C), using various reagents that can influence the structure and chemical composition and target physical properties of aerogels.

One stage hydrothermal synthesis of aerogel from reduced graphene oxide (**Figure 4a–f**) was first proposed by Xu et al. [13]. The work shows that the aerogel obtained by this method is mechanically strong (after 12 hours of hydrothermal treatment at T = 180°C, the sample demonstrated an elastic modulus of 290 ± 20 kPa), could support 100 g of weight with little deformation, and has good electrical conductivity (4.9 ± 0.2 mS/cm), thermal stability, and high specific capacity (160 ± 5 F/g). It should be noted that the initial concentration of hydrogel

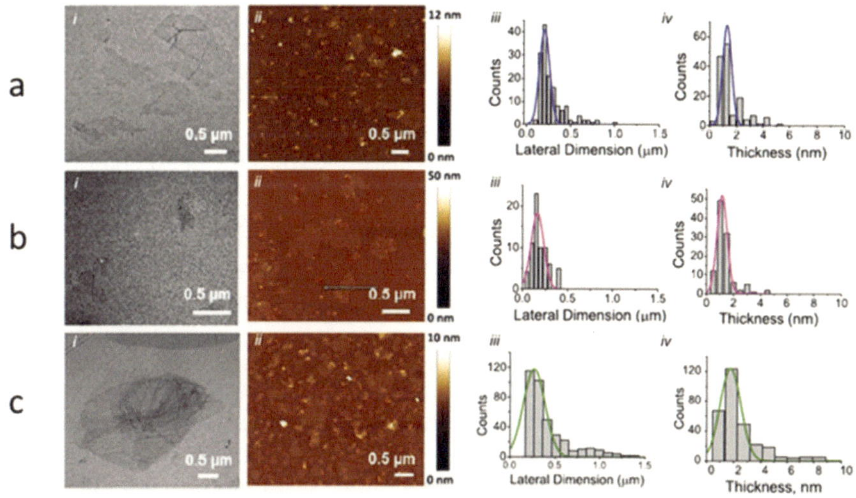

d

Sample	% π–π*	% O–C=O	% C=O	% C–O–C	% -OH	% C–C and C=C
GO-f	5.4 ± 1.4	9.5 ± 3.3	22 ± 6.7	19.0 ± 6.4	—	44.1 ± 3.8
GO-g	3.2 ± 1.1	5.6 ± 2.6	15.1 ± 3.5	31.5 ± 4.6	—	44.5 ± 7.3
GO-p	3.9 ± 0.6	8.2 ± 2.8	13.9 ± 4.9	30.0 ± 4.1	—	43.9 ± 0.7

Figure 3. Structural characterization of (a) GO-f, (b) GO-g, and (c) GO-p. TEM analysis (i), AFM height imaging (ii), lateral size distribution (iii), and thickness distribution (iv). The analysis in (iii) and (iv) was based on counting approximately 100 sheets captured in several AFM images. (d) Quantification of π—π*, carboxylic groups (O—C—O), carbonyls (C—O), epoxides (C—O—C), hydroxyls (C—OH), and graphitic structure (C—C and C—C) by high resolution C1s XPS spectra for GO-f, GO-g, and GO-p [12]. Three different GO thin sheets were synthesized from three starting graphite material: flakes (GO-f), ground (GO-g), and powder (GO-p).

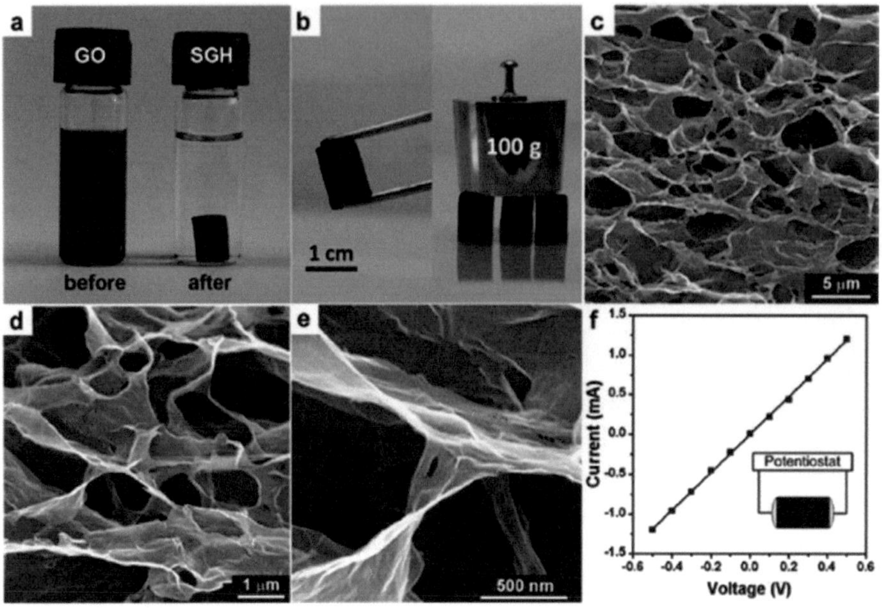

Figure 4. (a) Digital photographs of a homogeneous GO aqueous dispersion with a 2 mg/mL concentration before and after hydrothermal reduction at 180°C for 12 h; (b) photos of a strong self-assembled graphene hydrogel; (c–e) SEM images with different magnifications of the strong self-assembled graphene hydrogel interior microstructures; and (f) room temperature *I-V* curve of the strong self-assembled graphene hydrogel exhibiting Ohmic characteristic by the two-probe method for the conductivity measurements [13].

oxide graphene was only 0.5–2 mg/ml, which was enough to the self-gelation of GO dispersion. It is also shown that the concentration of dispersion of graphene oxide is an important factor affecting the process of aerogel structure formation and its final properties. A study of the dependence of rGO aerogel properties on the time of hydrothermal reaction showed that with increasing thermal treatment time both mechanical and electrophysical characteristics of aerogel are improved.

Using the features of a closed environment allows the in-situ modification of rGO aerogels [14]. In his work, a method for forming a three-dimensional structure of reduced graphene oxide with noble metals was developed by hydrothermal reaction of a suspension of graphene oxide containing noble metal salts and glucose. The obtained material, containing Pt in the structure (**Figure 5d–f**), demonstrated high catalytic activity. Other authors have shown that using a hydrothermal reduction process for graphene oxide in the presence of divalent metal ions (Ni^{2+}, Ca^{2+}, Co^{2+}), an aerogel of rGO decorated with metallic nanoparticles (**Figure 5a–c**) can be obtained in-situ [15]. A lot of work is devoted to the processes of decorating the particles of reduced graphene oxide in aerogels with various metal oxides. In particular, the great attention of researchers is attracted to Fe_3O_4 [16–18]. Special attention is also paid to the process of doping rGO aerogels with nitrogen. Doping with nitrogen by introducing into the system organic amines, ammonia, amino acids, and other nitrogen-containing compounds, at the stage of hydrothermal treatment, leads to a

http://dx.doi.org/10.5772/intechopen.78987

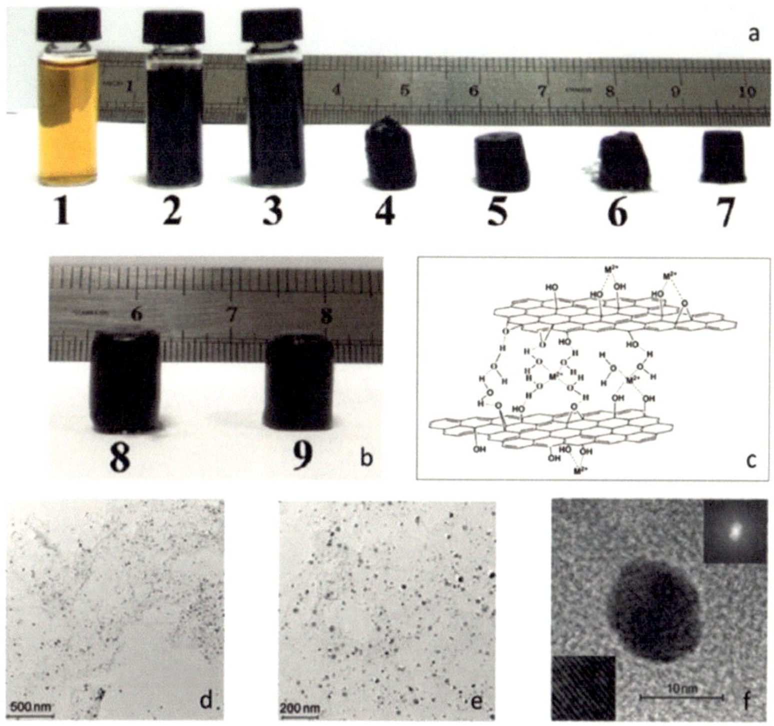

Figure 5. (a) Digital photographs of the GO before and after hydrothermal treatment: (1) GO, (2) rGO, and (3) rGO mixed with Ca^{2+} (m_{Ca}/m_{GO} is 0.003) suspensions as well as the gel-like rGO cylinders assembled by Ca^{2+}. Various m_{Ca}/m_{GO} were used: (4) 0.005, (5) 0.010, (6) 0.050, and (7) 0.100. (b) Photographs of the gel-like rGO samples assembled by (8) Ni^{2+} and (9) Co^{2+} with metal-ion/GO weight ratio of 0.010 [15]. (c) Schematic illustration of gel formation of rGO with divalent ion (M2+) linkage. (d–f) TEM images of graphene oxide decorated with Pd nanoparticles [14].

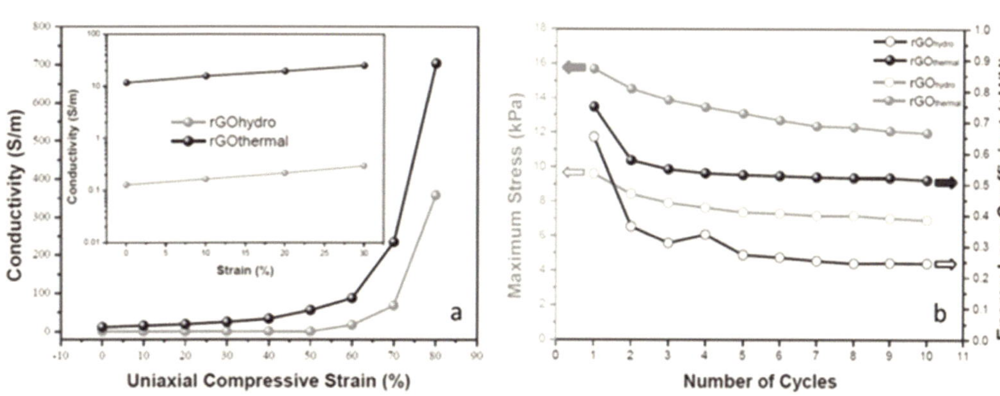

Figure 6. (a) Electrical conductivity of rGOhydro (prepared by hydrothermal synthesis) and rGOthermal (prepared by thermal annealing) aerogels when compressed along the axial direction. (b) Maximum stress (left pointing arrows) and energy loss coefficient (right pointing arrows) of rGO thermal aerogel during 10 cycles [23].

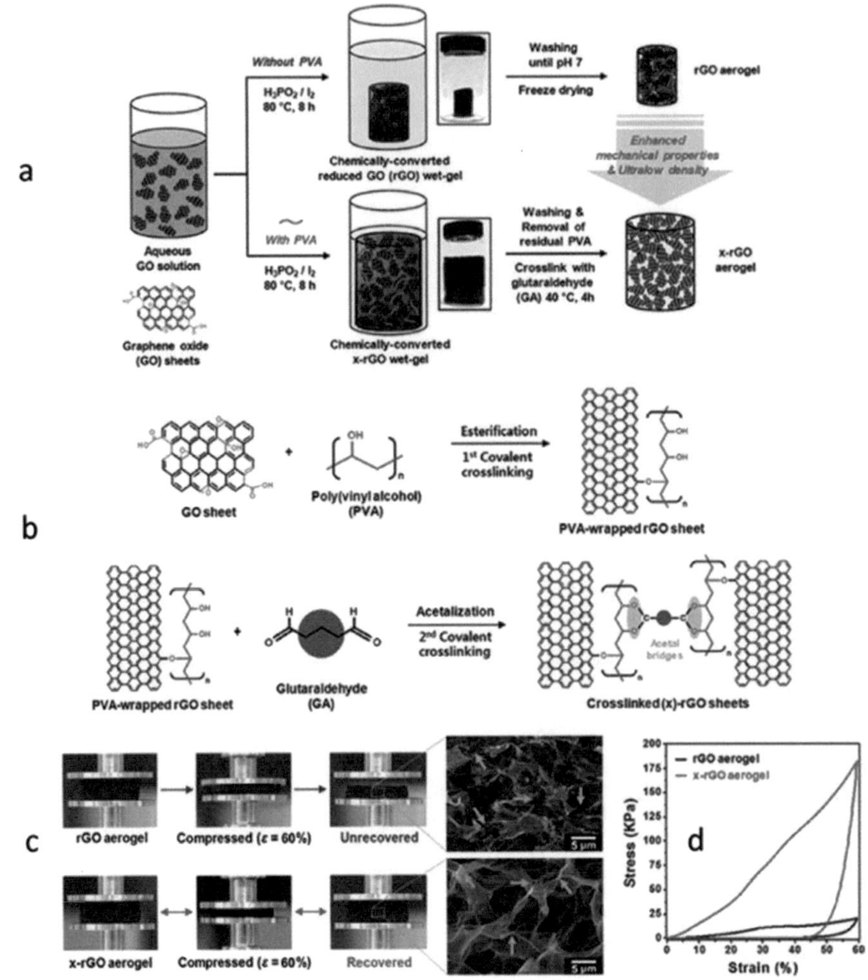

Figure 7. (a) Illustration of the different steps for fabricating the rGO (chemically converted reduced GO) aerogel and x-rGO (cross-linked rGO) aerogel. Insets at center: Digital photographs depict the as-prepared chemically converted rGO wet-gel and x-rGO wet-gel (top and bottom, respectively) after the self-assembly process. (b) A possible cross-linking mechanism between PVA-wrapped rGO sheets and Glutaraldehyde. PVA-bonded rGO sheets were covalently cross-linked by an acetal oxygen bridge through esterification reaction. (c) Digital photographs and SEM images (from left to right) of rGO and x-rGO aerogels after compression with 60% strain. Arrows indicate the material areas of deformed and recovered after the compression for rGO and x-rGO aerogels, respectively. (d) Compressive stress-strain curves were plotted with 60% strain for rGO and x-rGO [25].

significant increase in the electrophysical characteristics of aerogels from reduced graphene oxide [19–21], improves electrocatalytic properties [22], and also contributes to obtaining a more regular structure of the aerogel itself.

Moon et al. developed highly elastic and conductive N-doped monolithic graphene aerogels, using hexamethylenetetramine as a reducing agent, a nitrogen source, and a dispersion stabilizer of reduced graphene oxide [23]. To produce this material, hydrothermal synthesis was used, followed by annealing at T = 1000°C. The developed material showed good mechanical properties

http://dx.doi.org/10.5772/intechopen.78987

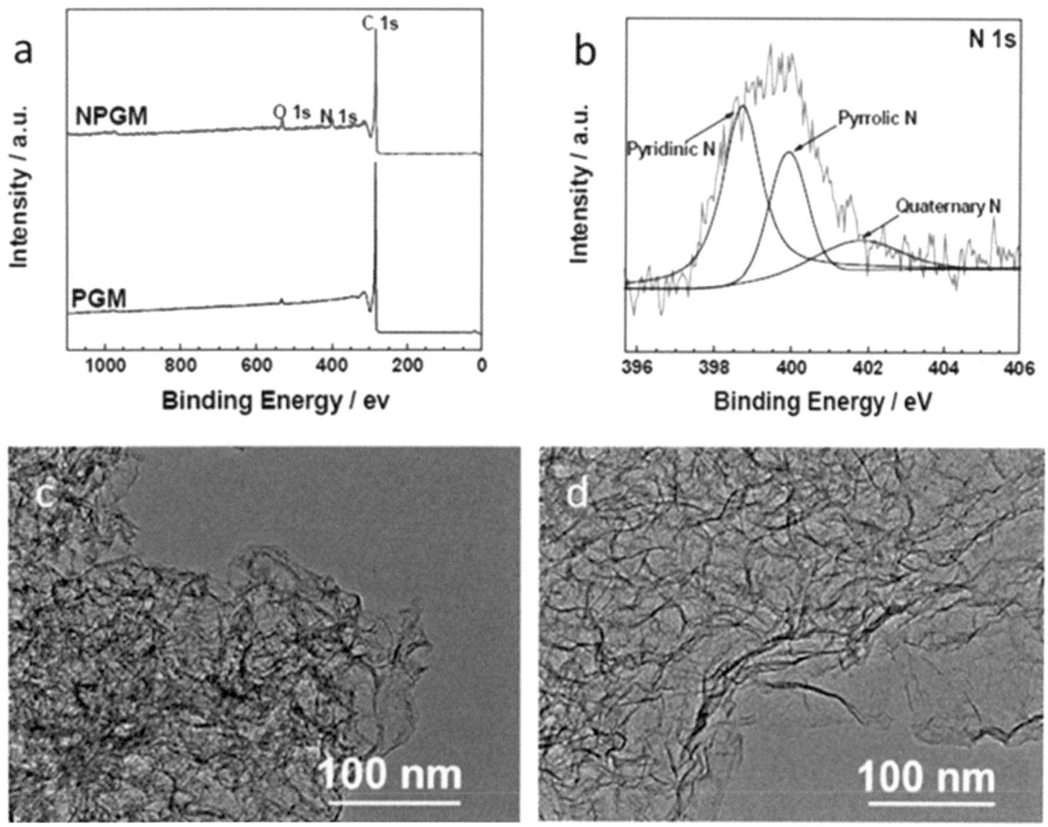

Figure 8. (a) XPS full spectrum of NPGM (nitrogen-doped porous graphene material) and PGM (porous graphene material). (b) N1s spectrum of NPGM. TEM images of NPGM (c) and PGM (d) [27].

(**Figure 6b**) and record electrical conductivity of 704 S/m for aerogels based on the reduced graphene oxide (**Figure 6a**). It has also been reported that ammonia is an effective agent for the production of nitrogen-doped aerogels [24]. The authors showed that when ammonia is introduced into the hydrothermal synthesis reactor, the degree of doping of the reduced graphene oxide is sufficiently high (8.4 atomic %) and the material has a high specific surface area of 830 m^2/g.

2.3. Polymer-assisted aerogel formation

Polymer-assisted aerogel formation with a cross-linking approach consists of the use of polymer components capable of binding to monolayers of graphene oxide, creating steric hindrances for stacking layers into stacks and reducing the available surface. The first version of this approach is the chemical bonding of polymer chains to graphene oxide particles. Reversibly deformable, highly elastic and strong aerogels based on reduced graphene oxide [25] have been developed using this method (**Figure 7**). Poly(vinyl alcohol) and glutaraldehyde were used as cross-linking components. The scheme of the processes is shown in the figure. Hypophosphorous acid and iodine were used as reducing agents. It is important to note that in addition to the high porosity of 92.16% and the low density of 10.6 mg/cm^3, the

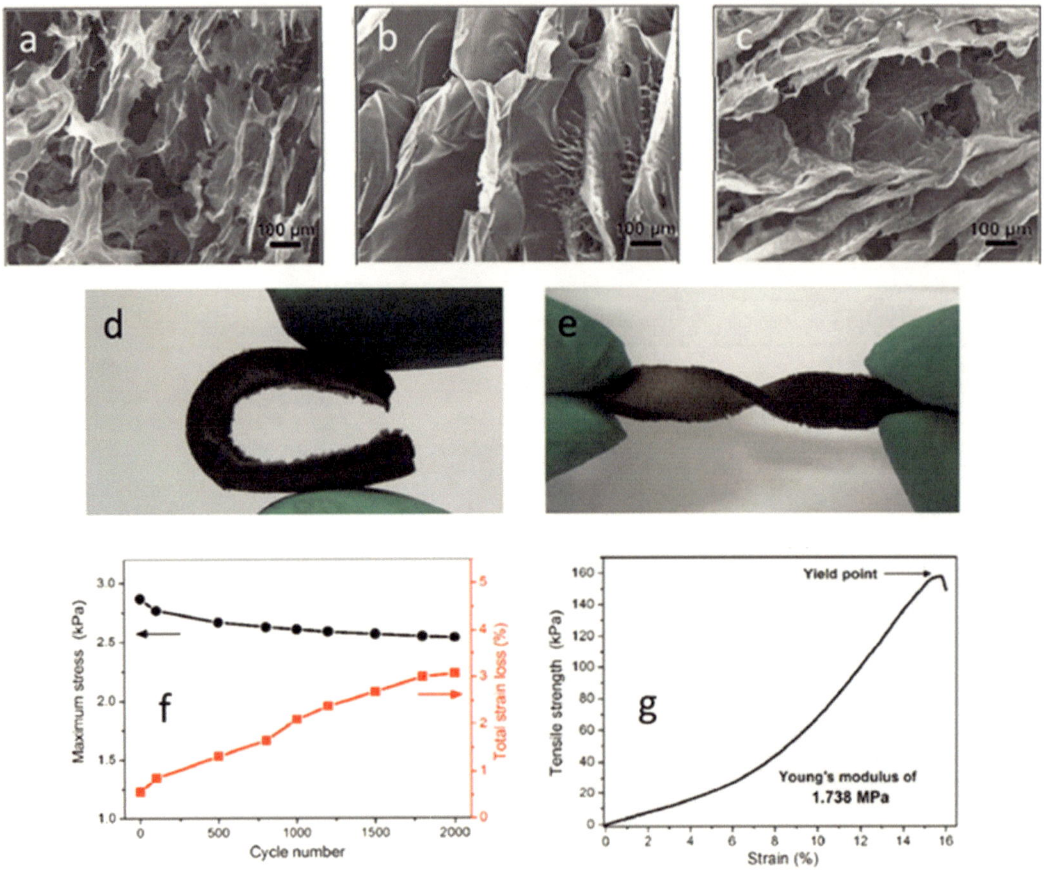

Figure 9. (a–c) SEM image of rGO aerogel, PI monolith, and rGO/PI nanocomposite, respectively. (d, e) Digital images show the high-level deformation of bend and torsion of rGO/PI nanocomposite. (f) Retention of maximum stress at 50% strain and total loss during 2000 cycles. (g) Tensile σ-ε curve for the rGO/PI [30].

combination of polymer chains and graphene oxide layers led to the mechanical strength of the material, whose structure was not fundamentally changed even after deformation by 60% (**Figure 7c**). Therefore, x-rGO aerogel exhibits 8.6 times higher compressive stress as compared with rGO aerogel (**Figure 7d**).

The second cross-linking option is the use of sol-gel technology, which is the preparation of a sol followed by its conversion to a gel-colloid system consisting of a liquid dispersion medium enclosed in a spatial grid formed by the connected particles of the dispersed phase. For the first time to produce aerogels based on reduced graphene oxide, this technology was applied by Worsley et al. [26]. The authors proposed the use of polymerization of resorcinol and formaldehyde in the presence of sodium carbonate in an aqueous dispersion of graphene oxide. The obtained material showed an increased electrical conductivity (~102 S/m) compared to the reduced graphene oxide (~0.5 S/m), as well as a high specific surface area of 584 m^2/g. Later Sui et al. also obtained an rGO aerogel with a high degree of nitrogen doping (5.8 atomic %), having a surface area of 1170 m^2/g [27] by sol-gel technology (**Figure 8**). The material was synthesized by freeze-drying an rGO/melamine-formaldehyde hydrogel and subsequent thermal treatment.

A third option for the formation of cross-linked aerogels is the polymerization of monomers in-situ in the presence of a dispersion of graphene oxide. Various polymers such as pyrrole [28] and aniline [29] have been used for this approach. Zhao et al. proposed a unique approach to the production of aerogel by hydrothermal reaction of graphene oxide with pyrrole, followed by electrochemical pyrrol polymerization [30]. The material showed excellent resistance to high loads without significant structural deformation and loss of elasticity. Qin et al. have developed a unique superelastic aerogel consisting of reduced graphene oxide and polyimide [30]. For its preparation, the introduction of a water-soluble polyamido acid into a dispersion of graphene oxide followed by lyophilization and thermal annealing was used. This aerogel showed extremely low density, excellent flexibility, the possibility of multiple reversible deformations (even after 2000 cycles), and good electrical conductivity (**Figure 9**). Li et al. proposed a utilization of in-situ polymerization of acrylamide to create rigid 3D structures based on reduced graphene oxide [31].

3. Applications of reduced graphene oxide aerogels

The first and the most developing direction of applications is the use of such aerogels as active electrode materials for supercapacitors. The supercapacitor is an electrochemical device for storage of electric energy on the surface of highly porous materials with an organic or inorganic electrolyte. At the heart of the work of supercapacitors, there are two processes—the formation of a double electrical layer at the material/electrolyte boundaries and electrochemical reactions on the surface of the electrode material, leading to the appearance of pseudocapacitance [32]. Both processes occur on the surface of the material during the charge/discharge of the device, so the energy capacitance of these devices is highly dependent on the surface area of the aerogels used. The combination of an electrically conductive three-dimensionally connected structure and good electrical conductivity of graphene-like materials makes them extremely attractive for this application. The high specific surface area of aerogels based on the reduced graphene oxide provides high capacity on a double electrical layer. To introduce a pseudocapacitive component, the aerogel surface is ordinarily decorated with transition metal oxides (Mn, V, etc.) capable of participating in redox reactions during charge/discharge of the device, making a significant contribution to the overall capacitance value [33–36]. Also interesting is the direction in the creation of flexible supercapacitors. In the framework of this direction, in addition to the reduced graphene oxide, various polymers are introduced into the aerogel, such as glucose [37], polyvinyl alcohol [25], polyaniline [38], etc.

The second, but not less interesting and intensively developing, direction is the use of aerogels based on reduced graphene oxide for rechargeable lithium-ion batteries. Graphene-like materials are the most widespread anode materials in commercial Li-ion batteries [39, 40]. The main role of rGO aerogel is to facilitate the multidimensional electronic transport routes and to reduce transport spaces between the electrode and the electrolyte. The consequence of this is an increase in the performance of the batteries and their cyclic stability. Sometimes, rGO aerogels containing metal, metal oxide, and metal sulfide are used as hybrid materials for the cathode of Li-ion batteries [41]. Similar structures containing Fe_3O_4 [42, 43] and Fe_2O_3 [39] show promising capacities (900–1100 $mA*h*g^{-1}$) with good cyclicality. SnO_2 is also used as an integral part of the FOG aerogel for this application [44]. Batteries with this material also show high performance (600–1200 $mA*h*g^{-1}$) [45].

Three-dimensional electroconductive structures of rGO aerogels are an excellent platform for creating electrochemical sensors, strain gauge sensor, and biosensors. Introduction to the structure of metals, oxides, and hydroxides of metals provides high sensitivity and electrochemical stability [46]. Ultraelastic aerogels based on rGO and carbon nanotubes were fabricated for use in a strain gauge sensor with adjustable voltage/pressure measurement [47]. The sensitivity is adjusted by changing the aerogel density. In the compression test, the measurement coefficient was 230 and 125% for deformations of 30 and 60%, respectively. The aerogel containing gold nanoparticles in its structure was used for the electrochemical determination of hydroquinone and o-dihydroxybenzene [48]. The detection limit is 1.5×10^{-8} M for hydroquinone and 3.3×10^{-9} M for o-dihydroxybenzene.

In the technique, actuators are transducers that convert an input signal (electrical, optical, mechanical, pneumatic, etc.) into an output signal (usually in motion) that acts on the control object. Devices of this type include electric motors; electric, pneumatic, or hydraulic actuators; relay devices; comb drives; DMD mirrors; electroactive polymers; robotic grasping mechanisms, drives for their moving parts, including solenoid actuators and voice coils; and many others. Recently, the actuators have been intensively studied as potential devices in flexible displays, soft robotics, and haptic devices. rGO-based aerogels are ideal candidates for such devices, because they have high porosity, are ultra-light, flexible, and resilient. To be able to act on the aerogel with magnetic forces, magnetic nanoparticles of Fe_3O_4 were introduced into it [49]. This material demonstrated great magnetic field-induced actuations of 52 and 35% along the radial and axial directions, respectively. Also, several works on actuators based on materials with shape memory are known. Li et al. developed an actuator based on an aerogel from rGO and trans-1,4-polyisoprene, which showed a strain of 80% at 10 V [31].

4. Conclusion

Aerogels based on reduced graphene oxide are promising materials and attracting the interest of many researchers due to unique physicochemical properties. High specific surface area, extremely low density, high porosity, uniqueness of structure, and good electrical conductivity make these materials indispensable in many applications. Main researches are carried out in the direction of surface modification with various materials in order to improve the mechanical, electrophysical, and structural properties of these materials, and the variety and number of articles in this field testify to the incredible promise of materials based on graphene-like particles. However, there are a number of problems that need to be overcome in order to bring most of the developments beyond the scope of laboratory research. First, it is necessary to develop simple methods for obtaining regular structures based on reduced graphene oxide, suitable for real use. Secondly, special attention needs to be given to a detailed study of the mechanisms of the structure formation of such materials, since at the moment many of the processes are only described, but not explained theoretically. Third, one of the main factors is the high cost of graphene-like materials, which means that efforts should be made to develop cheaper methods for the synthesis of GO and rGO. However, despite the fact that in the near future the researchers will have to solve a number of the problems described above, one can say unambiguously that materials based on graphene-like particles are among the most promising for revolutionary changes in technology, science, and the life of all mankind.

Acknowledgements

This work was partially supported by Russian Foundation for Basic Research (project 16-29-06440).

Conflict of interest

The authors declare that the research was conducted in the absence of any commercial or financial relationships that could be construed as a potential conflict of interest.

Author details

Gudkov Maksim Vladimirovich* and Valery Pavlovich Melnikov

*Address all correspondence to: gudkovmv@gmail.com

Semenov Institute of Chemical Physics of Russian Academy of Sciences, Moscow, Russia

References

[1] Zhang X, Sui Z, Xu B, Yue S, Luo Y, Zhan W, et al. Mechanically strong and highly conductive graphene aerogel and its use as electrodes for electrochemical power sources. Journal of Materials Chemistry. 2011;**21**:6494-6497. DOI: 10.1039/C1JM10239G

[2] Chen W, Yan L. In situ self-assembly of mild chemical reduction graphene for three-dimensional architectures. Nanoscale. 2011;**3**:3132-3137. DOI: 10.1039/C1NR10355E

[3] Wu ZS, Winter A, Chen L, Sun Y, Turchanin A, Feng X, et al. Three-dimensional nitrogen and boron co-doped graphene for high-performance all-solid-state supercapacitors. Advanced Materials. 2012;**24**:5130-5135. DOI: 10.1002/adma.201201948

[4] Zhang L, Chen G, Hedhili MN, Zhang H, Wang P. Three-dimensional assemblies of graphene prepared by a novel chemical reduction-induced self-assembly method. Nanoscale. 2012;**4**:7038-7045. DOI: 10.1039/C2NR32157B

[5] Yang S, Zhang L, Yang Q, Zhang Z, Chen B, Lv P, et al. Graphene aerogel prepared by thermal evaporation of graphene oxide suspension containing sodium bicarbonate. Journal of Materials Chemistry A. 2015;**3**:7950-7958. DOI: 10.1039/C5TA01222H

[6] Si W, Wu X, Zhou J, Guo F, Zhuo S, Cui H, et al. Reduced graphene oxide aerogel with high-rate supercapacitive performance in aqueous electrolytes. Nanoscale Research Letters. 2013;**8**:247. DOI: 10.1186/1556-276X-8-247

[7] Gudkov MV, Gorenberg AY, Shchegolikhin AN, Shashkin DP, Mel'nikov VP. Explosive reduction of graphite oxide by hydrazine vapor at room temperature. Doklady Physical Chemistry. 2018;**478**(1):11-14. DOI: 10.1134/S0012501618010037

[8] Li D, Müller MB, Gilje S, Kaner RB, Wallace GG. Processable aqueous dispersions of graphene nanosheets. Nature Nanotechnology. 2008;**3**:101-105. DOI: 10.1038/nnano.2007.451

[9] Chabot V, Higgins D, Yu A, Xiao X, Chen Z, Zhang J. A review of graphene and graphene oxide sponge: Material synthesis and applications to energy and the environment. Energy & Environmental Science. 2014;**7**:1564-1596. DOI: 10.1039/c3ee43385d

[10] Qin SY, Liu XJ, Zhuo RX, Zhang XZ. Microstructure-controllable graphene oxide hydrogel film based on a pH-responsive graphene oxide hydrogel. Macromolecular Chemistry and Physics. 2012;**213**:2044-2051. DOI: 10.1002/macp.201200281

[11] Bai H, Li C, Wang X, Shi G. On the gelation of Graphene oxide. Journal of Physical Chemistry C. 2011;**115**:5545-5551. DOI: 10.1021/jp1120299

[12] Jasim D, Lozano N, Kostarelos K. Synthesis of few-layered, high-purity graphene oxide sheets from different graphite sources for biology. 2D Materials. 2016;**3**:014006. DOI: 10.1088/2053-1583/3/1/014006

[13] Xu Y, Sheng K, Li C, Shi G. Self-assembled graphene hydrogel via a one-step hydrothermal process. ACS Nano. 2010;**4**:4324-4330. DOI: 10.1021/nn101187z

[14] Tang Z, Shen S, Zhuang J, Wang X. Noble-metal-promoted three-dimensional macroassembly of single-layered graphene oxide. Angewandte Chemie, International Edition. 2010;**49**:4603-4607. DOI: 10.1002/anie.201000270

[15] Jiang X, Ma Y, Li J, Fan Q, Huang W. Self-assembly of reduced graphene oxide into three-dimensional architecture by divalent ion linkage. Journal of Physical Chemistry C. 2010;**114**:22462-22465. DOI: 10.1021/jp108081g

[16] Cong HP, Ren XC, Wang P, Yu SH. Macroscopic multifunctional graphene-based hydrogels and aerogels by a metal ion induced self-assembly process. ACS Nano. 2012;**6**:2693-2703. DOI: 10.1021/nn300082k

[17] Wu ZS, Yang S, Sun Y, Parvez K, Feng X, Müllen K. 3D nitrogen-doped graphene aerogel-supported Fe3O4 nanoparticles as efficient electrocatalysts for the oxygen reduction

reaction. Journal of the American Chemical Society. 2012;**134**:9082-9085. DOI: 10.1021/ja3030565

[18] Xu X, Li H, Zhang Q, Hu H, Zhao Z, Li J, et al. Self-sensing, ultralight, and conductive 3D graphene/iron oxide aerogel elastomer deformable in a magnetic field. ACS Nano. 2015;**9**:3969-3977. DOI: 10.1021/nn507426u

[19] Chen P, Yang JJ, Li SS, Wang Z, Xiao TY, Qian YH, et al. Hydrothermal synthesis of macroscopic nitrogen-doped graphene hydrogels for ultrafast supercapacitor. Nano Energy. 2013;**2**:249-256. DOI: 10.1016/j.nanoen.2012.09.003

[20] Leea JW, Kob JM, Kim J-D. Hydrothermal preparation of nitrogen-doped graphene sheets via hexamethylenetetramine for application as supercapacitor electrodes. Electrochimica Acta. 2012;**85**:459-466. DOI: 10.1016/j.electacta.2012.08.070

[21] Wang T, Wang L, Wu D, Xia W, Zhaoa H, Jia D. Hydrothermal synthesis of nitrogen-doped graphene hydrogels using amino acids with different acidities as doping agents. Journal of Materials Chemistry. A. 2014;**2**:8352. DOI: 10.1039/c4ta00170b

[22] Suhaga D, Singhb A, Chattopadhyayc S, Chakrabartid S, Mukherjee M. Hydrothermal synthesis of nitrogen doped graphene nanosheets from carbon nanosheets with enhanced electrocatalytic properties. RSC Advances. 2015;**5**(50):39705-39713

[23] Moon IK, Yoon S, Chun KY, Oh J. Highly elastic and conductive N-doped monolithic graphene aerogels for multifunctional applications. Advanced Functional Materials. 2015;**25**:6976-6984. DOI: 10.1002/adfm.201502395

[24] Sui ZY, Meng YN, Xiao PW, Zhao ZQ, Wei ZX, Han BH. Nitrogen-doped graphene aerogels as efficient supercapacitor electrodes and gas adsorbents. ACS Applied Materials & Interfaces. 2015;**7**:1431-1438. DOI: 10.1021/am5042065

[25] Hong JY, Bak BM, Wie JJ, Kong J, Park HS. Reversibly compressible, highly elastic, and durable graphene aerogels for energy storage devices under limiting conditions. Advanced Functional Materials. 2015;**25**:1053-1062. DOI: 10.1002/adfm.201403273

[26] Worsley MA, Pauzauskie PJ, Olson TY, Biener J, Satcher JH, Baumann TF. Synthesis of graphene aerogel with high electrical conductivity. Journal of the American Chemical Society. 2010;**132**:14067-14069. DOI: 10.1021/ja1072299

[27] Sui ZY, Wang C, Yang QS, Shu K, Liu YW, Han BH, et al. A highly nitrogen-doped porous graphene—An anode material for lithium ion batteries. Journal of Materials Chemistry A. 2015;**3**:18229-18237. DOI: 10.1039/C5TA05759K

[28] Zhao Y, Liu J, Hu Y, Cheng H, Hu C, Jiang C, et al. Highly compression-tolerant supercapacitor based on polypyrrole-mediated graphene foam electrodes. Advanced Materials. 2012;**25**:591-595. DOI: 10.1002/adma.201203578

[29] Tai Z, Yan X, Xue Q. Three-dimensional graphene/polyaniline composite hydrogel as supercapacitor electrode. Journal of the Electrochemical Society. 2012;**159**:A1702-A1709. DOI: 10.1149/2.058210jes

[30] Qin Y, Peng Q, Ding Y, Lin Z, Wang C, Li Y, et al. Lightweight, superelastic, and mechanically flexible graphene/polyimide nanocomposite foam for strain sensor application. ACS Nano. 2015;**9**:8933-8941. DOI: 10.1021/acsnano.5b02781

[31] Li C, Qiu L, Zhang B, Li D, Liu CY. Robust vacuum-/air-dried graphene aerogels and fast recoverable shape-memory hybrid foams. Advanced Materials. 2016;**28**:1510-1516. DOI: 10.1002/adma.201504317

[32] Zhai Y, Dou Y, Zhao D, Fulvio PF, Mayes RT, Dai S. Carbon materials for chemical capacitive energy storage. Advanced Materials. 2011;**23**:4828-4850. DOI: 10.1002/adma.201100984

[33] Liu Y, He D, Wu H, Duan J, Zhang Y. Hydrothermal self-assembly of manganese dioxide/manganese carbonate/reduced graphene oxide aerogel for asymmetric supercapacitors. Electrochimica Acta. 2015;**164**:154-162. DOI: 10.1016/j.electacta.2015.01.223

[34] Wu Y, Gao G, Wu G. Self-assembled three-dimensional hierarchical porous V2O5/graphene hybrid aerogels for supercapacitors with high energy density and long cycle life. Journal of Materials Chemistry. 2015;**A3**:1828-1832. DOI: 10.1039/C4TA05537C

[35] Yu G, Hu L, Vosgueritchian M, Wang H, Xie X, McDonough JR, et al. Solution-processed graphene/MnO2nanostructured textiles for high-performance electrochemical capacitors. Nano Letters. 2011;**11**:2905-2911. DOI: 10.1021/nl2013828

[36] He Y, Chen W, Li X, Zhang Z, Fu J, Zhao C, et al. Freestanding three-dimensional graphene/MnO2 composite networks as ultralight and flexible supercapacitor electrodes. ACS Nano. 2013;**7**:174-182. DOI: 10.1021/nn304833s

[37] Lee WSV, Peng E, Choy DC, Xue JM. Mechanically robust glucose strutted graphene aerogel paper as a flexible electrode. Journal of Materials Chemistry. 2015;**A3**:19144-19147. DOI: 10.1039/C5TA06072A

[38] Lv P, Tang X, Zheng R, Ma X, Yu K, Wei W. Graphene/polyaniline aerogel with superelasticity and high capacitance as highly compression-tolerant supercapacitor electrode. Nanoscale Research Letters. 2017;**12**:630. DOI: 10.1186/s11671-017-2395-z

[39] Xiao L, Wu D, Han S, Huang Y, Li S, He M, et al. Self-assembled Fe2O3/graphene aerogel with high lithium storage performance. ACS Applied Materials & Interfaces. 2013;**5**:3764-3769. DOI: 10.1021/am400387t

[40] Ren L, Hui KN, Hui KS, Liu Y, Qi X, Zhong J, et al. 3D hierarchical porous graphene aerogel with tunable meso-pores on graphene nanosheets for high-performance energy storage. Scientific Reports. 2015;**5**:14229. DOI: 10.1038/srep14229

[41] Zhou G, Li F, Cheng HM. Progress in flexible lithium batteries and future prospects. Energy & Environmental Science. 2014;**7**:1307-1338. DOI: 10.1039/C3EE43182G

[42] Chen W, Li S, Chen C, Yan L. Self-assembly and embedding of nanoparticles by in situ reduced graphene for preparation of a 3D graphene/nanoparticle aerogel. Advanced Materials. 2011;**23**:5679-5683. DOI: 10.1002/adma.201102838

[43] Wei W, Yang S, Zhou H, Lieberwirth I, Feng X, Müllen K. 3D graphene foams cross-linked with pre-encapsulated Fe3O4 nanospheres for enhanced lithium storage. Advanced Materials. 2013;**25**:2909-2914. DOI: 10.1002/adma.201300445

[44] Yao X, Guo G, Ma X, Zhao Y, Ang CY, Luo Z, et al. In situ integration of anisotropic SnO2 heterostructures inside three-dimensional graphene aerogel for enhanced lithium storage. ACS Applied Materials & Interfaces. 2015;**7**:26085-26093. DOI: 10.1021/acsami.5b07081

[45] Wang R, Xu C, Sun J, Gao L, Yao H. Solvothermal-induced 3D macroscopic SnO2/nitrogen-doped graphene aerogels for high capacity and long-life lithium storage. ACS Applied Materials & Interfaces. 2014;**6**:3427-3436. DOI: 10.1021/am405557c

[46] Yan X, Yao W, Hu L, Liu D, Wang C, Lee CS. Progress in the preparation and application of three-dimensional graphene-based porous nanocomposites. Nanoscale. 2015;**7**:5563-5577. DOI: 10.1039/C5NR00030K

[47] Lv P, Yu K, Tan X, Zheng R, Ni Y, Wang Z, et al. Super-elastic graphene/carbon nanotube aerogels and their applications as a strain-gauge sensor. RSC Advances. 2016;**6**:11256-11261. DOI: 10.1039/C5RA20342B

[48] Juanjuan Z, Ruiyi L, Zaijun L, Junkang L, Zhiguoa G, Guangli W. Synthesis of nitrogen-doped activated graphene aerogel/gold nanoparticles and its application for electrochemical detection of hydroquinone and o-dihydroxybenzene. Nanoscale. 2014;**6**:5458-5466. DOI: 10.1039/c4nr00005f

[49] Xu L, Xiao G, Chen C, Run L, Mai Y, Sun G, et al. Superhydrophobic and superoleophilic graphene aerogel prepared by facile chemical reduction. Journal of Materials Chemistry A. 2015;**3**:7498-7504. DOI: 10.1039/C5TA00383K

Chapter 5

Graphene Oxide-Based Biosensors

Lingwen Zeng, Shilin Cao, Hang Yin, Jun Xiong and Donghai Lin

Additional information is available at the end of the chapter

http://dx.doi.org/10.5772/intechopen.78222

Abstract

In this chapter, the latest developments in graphene oxide-based biosensors are presented. These biosensors are complexes of graphene oxide and biomacromolecules, including enzymes such as glucose oxidase, horseradish peroxidase, laccase, and nucleic acids such as DNA and RNA. The structure, design and preparation process (immobilization process) of the above graphene oxide-biomacromolecule composites were summarized. Some typical examples of immobilization of biological macromolecules are described. The immobilization efficiency and electrochemical performance of immobilized biomolecules based on graphene oxide were discussed, which may guide designing better graphene oxide-based biosensors.

Keywords: graphene oxide, biosensor, enzyme, nucleic acid

1. Introduction

Graphene is a new kind of two-dimensional single-atom carbon sheet with a single atom thick [1]. Nowadays, this so-called "thinnest in our universe" material [2] has attracted more and more attention, because of its unique properties such as unique electronic properties [3].

Graphene oxide (GO), one of the nanomaterials from graphene family, contains many reactive oxygen functional groups, such as hydroxyl group, a carboxyl group, an epoxy group [4]. It has been extensively used for biosensor research and application. In order to enhance the electrochemical properties of the GO-based biosensor, GO can be modified with other materials, such as macromolecules, small mass organic molecules, metallic oxide, and metallic/nonmetallic simple substances. In this chapter, we describe GO-based biosensors containing various composites of these materials with references such as GO-chitosan nanocomposites, GO-based glucose oxidase, chitosan-ferrocene/graphene oxide/glucose oxidase, metal oxides, HRP, multi nanomaterials, quantum dots, multiwall carbon nanotubes, DNA, miRNA, etc. We not only describe the relevant preparation process of the above biosensors but also introduce their electrochemical properties to provide more guidance for designing suitable GO-based biosensors.

2. Enzyme/graphene oxide based biosensor

2.1. GOD/graphene oxide (GO) based biosensor

Glucose oxidase (GOD) is an oxidoreductase, which can oxidize glucose to D-glucono-δ-lactone and form hydrogen peroxide. GOD has shown great potential in glucose biosensor, forage, medicine [5] and biocatalysis [6].

Improving the dispersion of GO ensures the efficient use of the GO-based biosensors. Chitosan is a biopolymer with unique physical/chemical properties and can be well soluble in aqueous acidic solution [7]. Kang et al. [7] firstly mixed graphene with chitosan solution to form a hybrid nanocomposite of graphene-chitosan. Then, this hybrid nanocomposite was coated onto the surface of a glassy carbon electrode (GCE). Finally, this electrode was incubated with GOD solution to form a GOD/graphene/chitosan sensor (**Figure 1**). The result showed that chitosan could improve the dispersion of the graphene and GOD enzyme molecules. The as-prepared GOD/graphene/chitosan sensor exhibited excellent sensitivity (37.93 $AmM^{-1}\ cm^{-2}$) and a much higher enzyme loading (1.12×10^{-9} mol cm^{-2}). Also, this biosensor could retain more than 95% of the enzyme activity after store of 7 days at 4°C.

Chitosan can be used not only as a stabilizing agent but also as reducing agent. Sun et al. [8] designed a graphene platelet-glucose oxidase (GP-GOD) biosensor (**Figure 2**). GO was dispersed in H_2O and mixed with 0.5 M chitosan solution. After stirring for 30 min at room temperature, this mixture was heated at 90°C for 2 h to form a graphene platelet composite (GP)

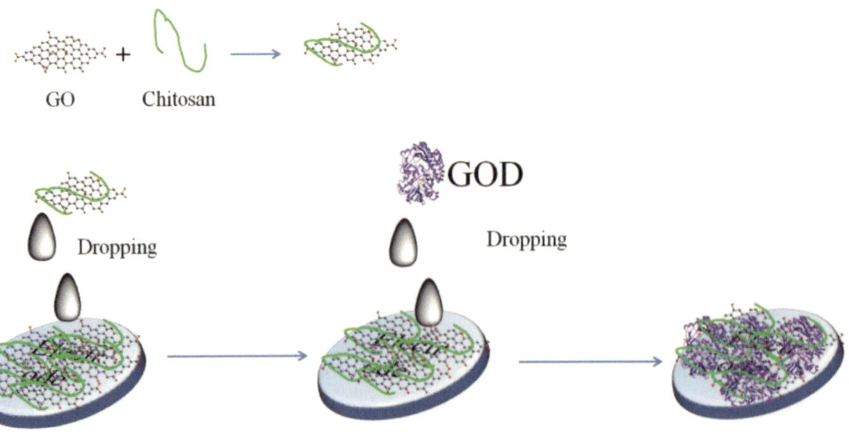

Figure 1. The scheme of GO-chitosan/GOD bioelectrode [7].

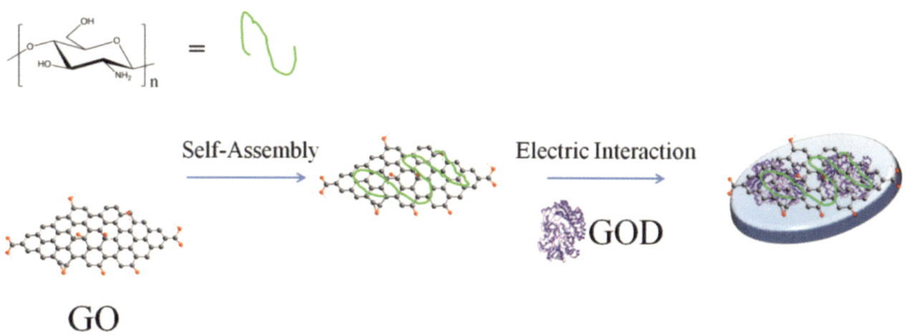

Figure 2. The scheme of GO-chitosan/GOD bioelectrode [8].

dispersion. Then, the GP dispersion was mixed with GOD solution under ultrasonic treatment and kept at 4°C for 12 h to obtain a large amount solid-state GP-GOD product. The glucose biosensor was constructed by deposition of the as-prepared GP-GOD product on a glassy carbon electrode. The linear relation against the concentration of glucose ranged from 2 to 22 mM (R = 0.9987) with an estimated detection limit to be 20 M at a signal-to-noise ratio (S/N) of 3.

Luo et al. [9] developed a GO-based glucose biosensor by a direct electro-deposition process. The graphene oxide, chitosan, and GOD are directly electrodeposited onto a glassy carbon electrode (GCE) by using electrochemical reduction under controlled direct electrical potential. This direct electro-deposition process is rapid (several minutes) and can produce uniform, controllable and reproducible films. The GO-chitosan-GOD composite was formed by dispersing GO (5 mg) in chitosan solution (0.2% w/v) and 5 mg mL^{-1} GOD added stepwise. The GCE was then immersed in the GO-chitosan-GOD solution while a fixed potential of −1.0 V was applied for 400 s. When the electrodeposition time increased from 100 to 400 s, the amount of the GOD entrapped in the film and their current response increased as well. However, while the electrodeposition time further increased from 400 to 900 s, the current response was not a significantly improved. The reason is that excessively thick films have negative effects on the GOD activity and prolong the response time. The as-prepared biosensor film indicated fast response (<3 s), a lower detection limit (0.4 M), and a linear range from 0.4 M to 2 mM towards glucose.

The direct electron transfer between electrode surface and active center of enzyme is commonly hindered. This is mainly because that the active center of the enzyme is buried in the globular structure of the protein molecule. To overcome this drawback, enzymes can be composited with conducting or redox polymers. Qiu et al. [10] designed a homogeneous chitosan-ferrocene/graphene oxide/glucose oxidase (CS-Fc/GO/GOD) nanocomposite film as a novel platform for glucose biosensor. The ferrocene branched chitosan (CS-Fc) was prepared by the following steps (**Figure 3**): (1) chitosan aqueous solution and ferrocenecarboxaldehyde (FcCHO) methanol solution was mixed at room temperature for 2 h to form the Schiff-base; (2) the $NaCNBH_3$ was added to the above mixture and CS-Fc. (3) the biosensor was constructed by covering the mixture onto the GCE and dried in air at room temperature (**Figure 4**). This CS-Fc/GO/GOD sensor exhibited a wide linear range, excellent sensitivity, good reproducibility, and long-term stability.

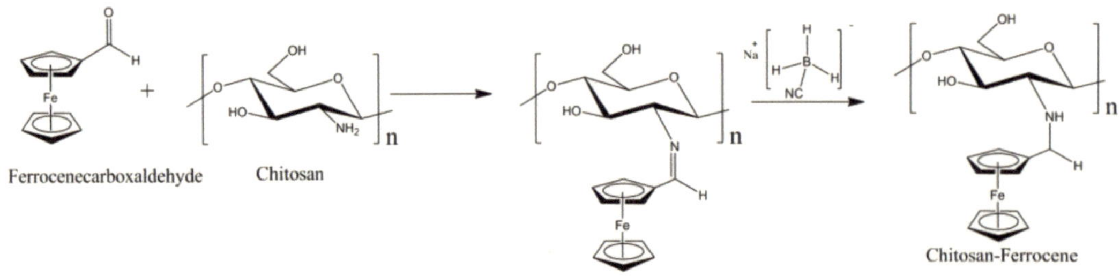

Figure 3. The preparation scheme of the ferrocene branched chitosan [10].

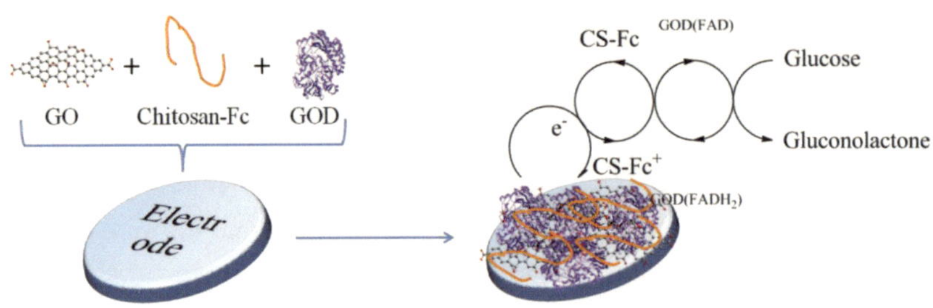

Figure 4. The preparation scheme of CS-Fc/GO/GOD-based glucose biosensor [10].

Apart from chitosan, metal oxides were also used to facilitate the immobilization of GOD onto GO to form biosensors. ZnO is a nontoxic material with good conductivity. It has a high isoelectric point at about 9.5. Thus, the electrostatic interaction between ZnO and GOD (with an isoelectric point at 4.2) can occur. Chen et al. [11] prepared ZnO-microflowers on reduced graphene oxide (RGO) modified GCE by using simple electrodeposition (**Figure 5**). This positively charged ZnO/RGO composite self-assembled with negatively charged GOD and fabricated an RGO/ZnO/GOD biosensor. The linear range of the biosensor was 0.02–6.24 mM with a detection limit of 0.02 mM and sensitivity of 18.97 μA mM^{-1}.

Carbon nanotubes are cylindrical nanostructural carbon allotropes with unique electronic, optical properties. Incorporation of carbon nanotubes into GO can enhance direct electron transfer in a biosensor. However, the carbon nanotubes are difficult to disperse homogeneously

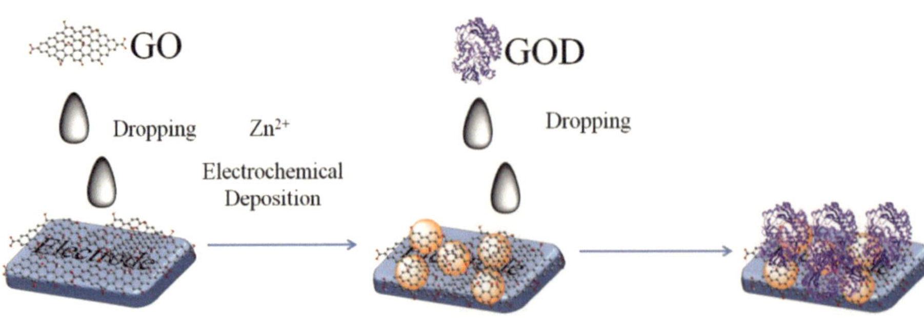

Figure 5. The preparation scheme of ZnO/GO/GOD-based glucose biosensor [11].

http://dx.doi.org/10.5772/intechopen.78222

in aqueous solution because of its hydrophobic surface and internal van der Waals interactions [12]. Surface modification of carbon nanotubes by GO is a way of solving this problem. Chen et al. [13] dispersed carbon nanotubes with GO aqueous homogeneous suspension to obtain stable carbon nanotubes/GO composite (**Figure 6**). The GOD was positively charged, and carbon nanotubes/GO composite is negatively charged. Thus, the GOD was immobilized by carbon nanotubes/GO composite through the electrostatic interaction as well as physical adsorption. The as-prepared biosensor is reproducible with enhanced direct electron transfer. The linear range of the biosensor was 0.1–19.82 mM with a detection limit of 0.028 mM.

2.2. Horseradish peroxidase/GO based biosensors

HRP is extensively used in clinical diagnosis. It can oxidize chromogenic substrates to colored products by using hydrogen peroxide [13]. The characteristic color change can be easily detected by spectrophotometric methods [14].

Combining GO sheets with chemiluminescence (CL) regents can facilitate the preparation of a sensitive sensor with attracting CL property. In the work of Liu et al. [15], N-aminobutyl-N-ethylisoluminol (ABEI) functionalized GO hybrids (ABEI-GO) was first synthesized by adding the ABEI alkaline solution into a stable GO suspension for 24 h at room temperature with stirring. HRP buffer solution was then mixed with the ABEI-GO suspension to form an ABEI-GO@HRP hybrid (**Figure 7**). In this strategy, there might be two assembly ways between HRP and ABEI-GO: (1) strong electrostatic interaction between HRP and GO, (2) interactions with hydrogen bonding. The results suggested an excellent CL properties for the detection of H_2O_2, exceeding those of previous reports. The ABEI-GO@HRP sensor showed a detection limit of 47 fM at physiological pH condition.

GO-based biosensors constructed with multi nanomaterials have also been investigated [16]. For example, many researchers have validated the remarkable electrocatalytic properties and biocompatibility of graphene-gold nanocomposite (G-AuNP). CdTe-CdS, the core-shell quantum dots, could significantly enhance the charge transfer, enabling nanosensor exploiting high intensity. Taken together, Gu and co-workers [16] fabricated a biosensor to detect hydrogen peroxide, integrating the benefits of G-AuNP, CdTe-CdS, and AuNPs (**Figure 8**). Such a biosensor was constructed by successively dropping casting G-AuNP, CdTe-CdS, AuNPs,

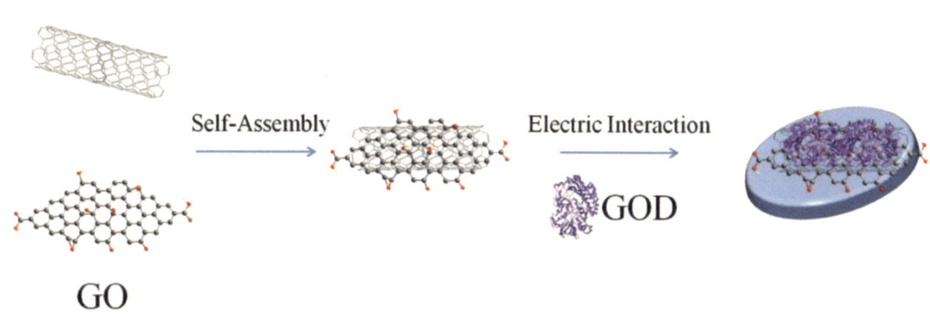

Figure 6. The preparation scheme of carbon nanotube/GO/GOD-based glucose biosensor [12].

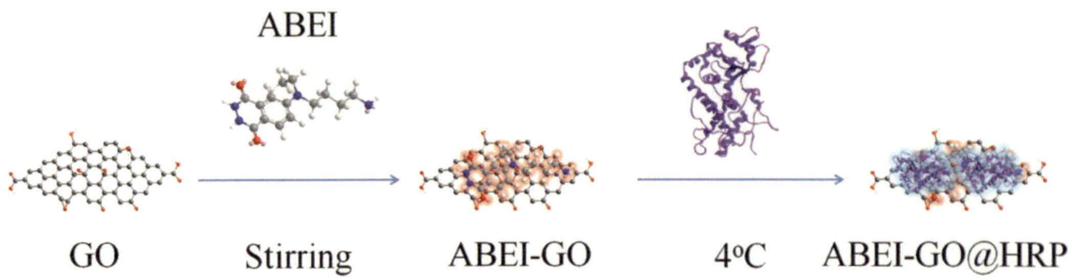

Figure 7. Scheme of the preparation process of the ABEI-GO@HRP composite [15].

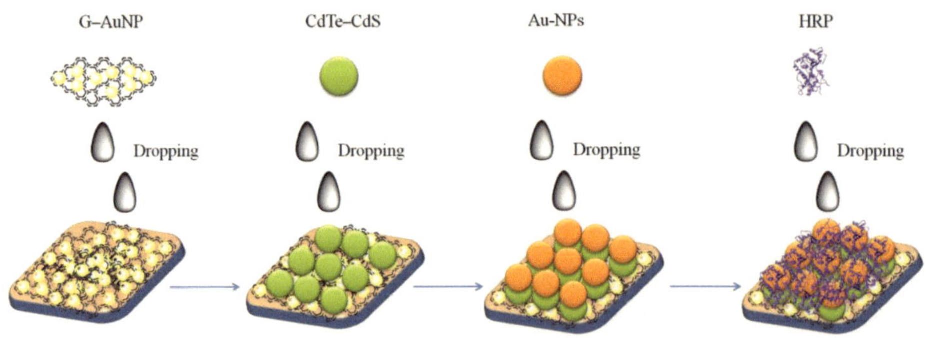

Figure 8. Scheme of the preparation process of the HRP/AuNPs/CdTe-CdS/G-AuNP/GE [16].

and HRP onto the surface of the gold electrode step by step. The Au-NPs were synthesized by citrate reduction in the presence of glutin [17] (**Figure 9**). The G-AuNPs were prepared by in-situ reduction of the $HAuCl_4$-loaded GO (**Figure 10**). The result exhibited that the biosensor displays an admirable sensitivity, low detection limit (S/N = 3) (3.2×10^{-11} M), wide calibration range (from 1×10^{-10} to 1.2×10^{-8} M) and good long-term stability (20 weeks).

Multiwall carbon nanotubes (MWNTs) have a poor solubility in water, which limits their application in biosensors. To overcome this obstacle, Zhang et al. [18] synthesized a well

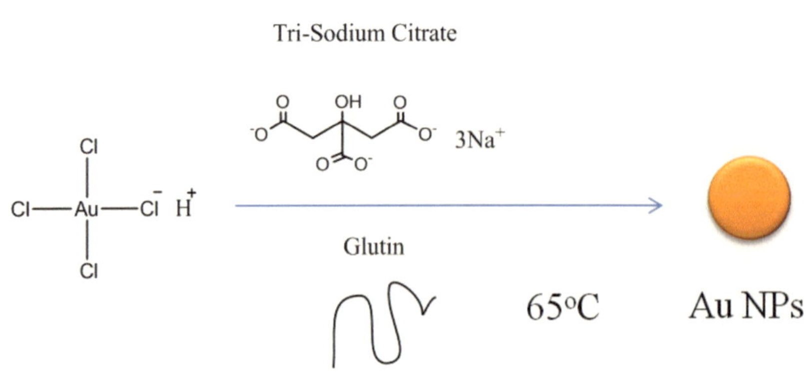

Figure 9. Scheme of the preparation process of the Au NPs by citrate reduction [17].

http://dx.doi.org/10.5772/intechopen.78222

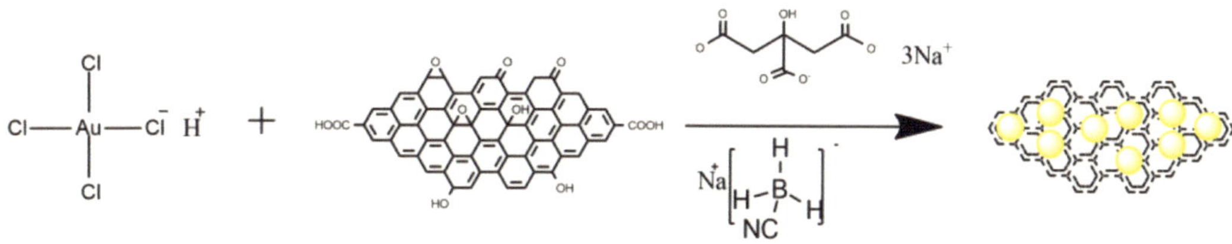

Figure 10. Scheme of the preparation process of the G-AuNPs were synthesized by in-situ reduction of the $HAuCl_4$-loaded GO [17].

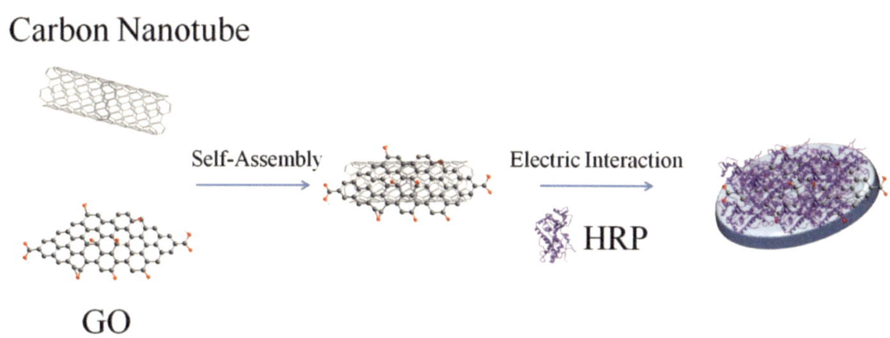

Figure 11. The preparation scheme of carbon nanotube/GO/GOD-based glucose biosensor [18].

depressed GO-MWNT hybrid nanomaterial aqueous solution that carried a negative charge. Subsequently, the as-prepared GO-MWNT aqueous solution was dropped onto the GC electrode, followed by adding HRP onto the GO-MWNT/GC (**Figure 11**). The result indicated that the direct electron transfer between immobilized enzyme and the GC electrode was enhanced by GO-MWNT composite. For detection of H_2O_2, the detection limit for the sensor was 1.17 μM on S/N-3, and the sensitivity of HRP/GO-MWNT/GC electrode was 563.7 mA cm^{-2} M^{-1}. For the reduction of $NaNO_2$, the sensitivity and the detection limit was 0.6 mA cm^{-2} M^{-1} and 12 mM (S/N = 3), respectively. Furthermore, this novel electrode showed excellent stability for less than 5% activity of 15 days.

Nafion, a commercial tetrafluoroethylene-perfluoro-3, 6-dioxa-4-methyl-7-octenesulfonic acid copolymer, has also been used to modify the GO electrode. In the work of Zhang et al. [19], Nafion solution was mixed with GO by ultrasonication, then HRP was added into the prepared mixed solution, following by casting onto the GCE. The as-prepared electrode, HRP/GO/Nafion/GCE, was proven to have a favorable electrocatalytic response with excellent linear relationships from 1.0 μM to 1.0 mM and the detection limits of 4.0×10^{-7} M (S/N = 3). Furthermore, the HRP/GO/Nafion/GCE biosensor showed satisfactory stability for less than 5% of reduced activity after 4 weeks of storage.

An excellent GO-based biosensor means to possess good electron-transfer property. Co-immobilizing Cytochrome *c* (Cyt *c*) and HRP on GO-chitosan nanocomposite were tried to fabricate a bi-protein electrode by Wan et al. [20]. Firstly, GO-chitosan nanocomposites were synthesized by stirring GO solution and chitosan solution. Then the HRP-Cyt *c*/GO-chitosan/Cyt *c*/MUA-MCH/Au electrode was produced by a layer-by-layer technique. It was found that the as-prepared biosensor can have an effective response to detecting H_2O_2 within 2 s, along with the linear range from 20 to 330 μM and detection limit of 6.68 μM (S/N = 3). Moreover, the electrode retained most of the activity for 2 weeks.

Owing to the good biocompatibility and large surface area of Co_3O_4 nanosheets (**Figure 12**), Co_3O_4 nanosheets can be used to enhance the electric transfer between enzyme and electrode in a biosensor. Herein, Liu et al. [21] firstly mixed Co_3O_4 suspension with HRP, following with addition of rGO (**Figure 13**). Then the mixture was dropped cast onto the GCE to form the Co_3O_4-HRP/rGO/GCE electrode successfully. This electrode held a higher HRP loading (with a concentration of 1.48×10^{-10} mol cm^{-2}) than that of monolayer coverage. Furthermore, the as-modified biosensor presented an excellent electronic response with a linear a range from 1 to 5400 μM, a limit of detection of 0.21 μM and a limit of quantification of 0.58 μM for detection of $NaNO_2$. Though the study of the stability of electrode was at 4°C for 4 weeks, they validated that electrode can hold 94.1% of its activity.

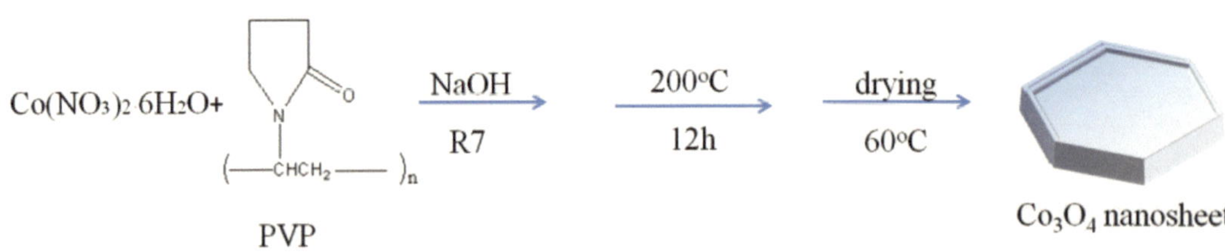

Figure 12. The preparation scheme of Co_3O_4 nanosheet [21].

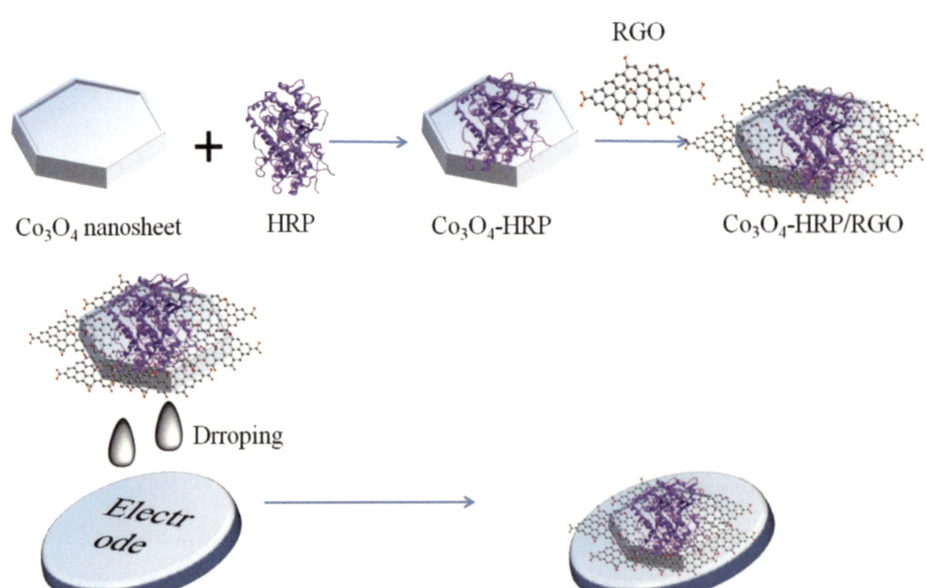

Figure 13. The preparation scheme of Co_3O_4 nanosheet/RGO/HRP biosensor [21].

http://dx.doi.org/10.5772/intechopen.78222

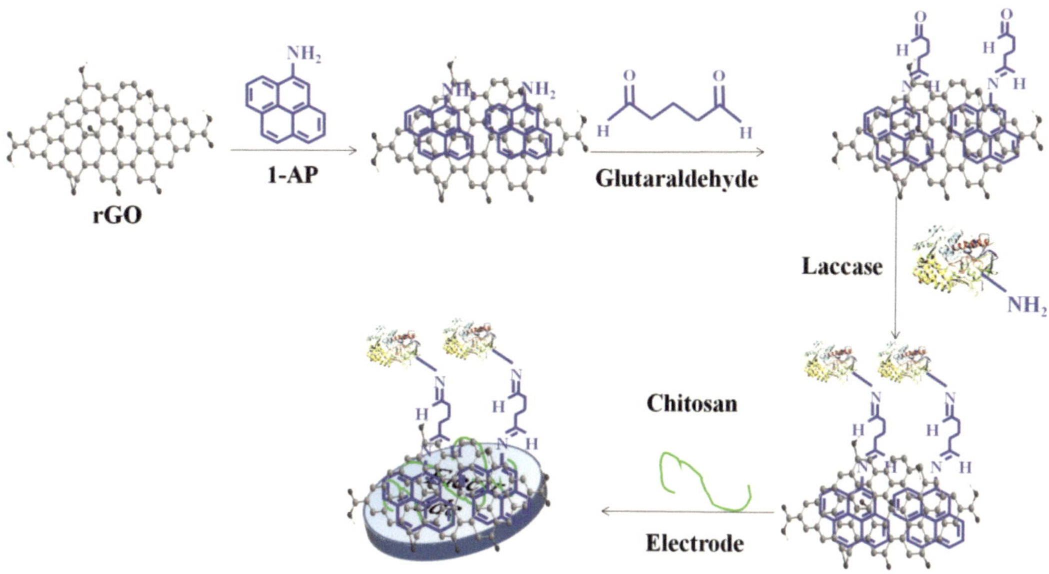

Figure 14. The preparation scheme of AP-rGOs/chitosan/laccase biosensor [23].

Palanisamy et al. [22] reported a novel electrode synthesis based on a screen-printed carbon electrode (SPCE) for the detection of H_2O_2. The electrode was fabricated by dropping GO-HRP composite onto SPCE to form ERGO sensor. The result showed that ERGO had an excellent enzyme loading, and the surface coverage concentration of HRP onto SPCE/ERGO-HRP was calculated to be 7.32×10^{-10} mol cm^{-2}. Moreover, the linear range of detection was 9–195 μM, and sensitivity of the sensor is 0.09 μA μM^{-1} cm^{-2}.

2.3. Laccase/graphene oxide (GO) based biosensor

Laccase is a kind of blue multi-copper oxidase and can catalyze phenols in the presence of oxygen. Thus, this enzyme can be used for the fabrication of phenols detection [23].

Zhou et al. [23] prepared a 1-aminopyrene-reduced graphene oxides (AP-rGOs) composite via the- interaction between the pyrenyl group of 1-aminopyrene and graphene (**Figure 14**). Then they covalently immobilized the laccase onto the AP-rGOs form Lac/AP-rGOs by using glutaraldehyde as cross-linker. After mixing chitosan with Lac/AP-rGOs, the Lac/AP-rGOs/chitosan stock solution was dropped onto GCE. The biosensor was used for the detection of phenols in water samples. The result showed that the biosensor exhibited a fast response time (<5 s), high stability (retained >97% activity after 7 days of storage).

3. Nucleic acids/graphene oxide-based biosensor

Previous literature indicated that the GO-chitosan composite could also be used for DNA biosensor fabrication [24]. The GO-chitosan electrode was activated by glutaraldehyde and covalently cross-linked with *Salmonella typhi* specific 5′-amine labeled single-stranded (ss)

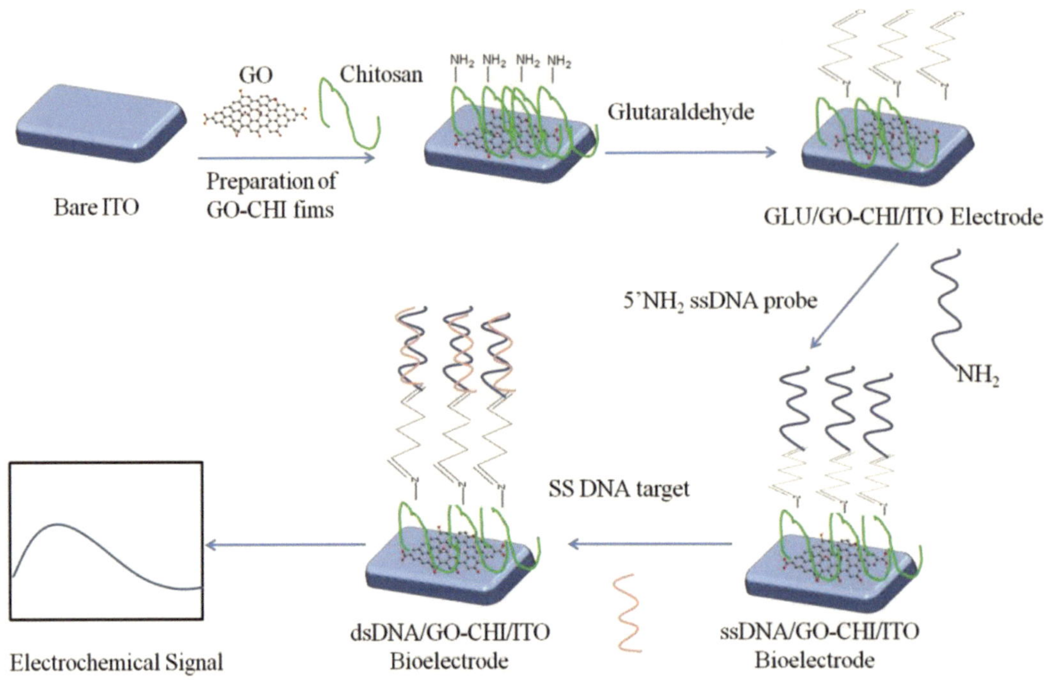

Figure 15. The scheme of ssDNA/GO-chitosan/ITO bioelectrode [24].

DNA probe (5′NH_2-ssDNA probe) (**Figure 15**). This DNA biosensor exhibited good ability to detect both complementary and non-complementary target. The linear range of detection was 10 fM–50 nM and the detection limit was 10 fM.

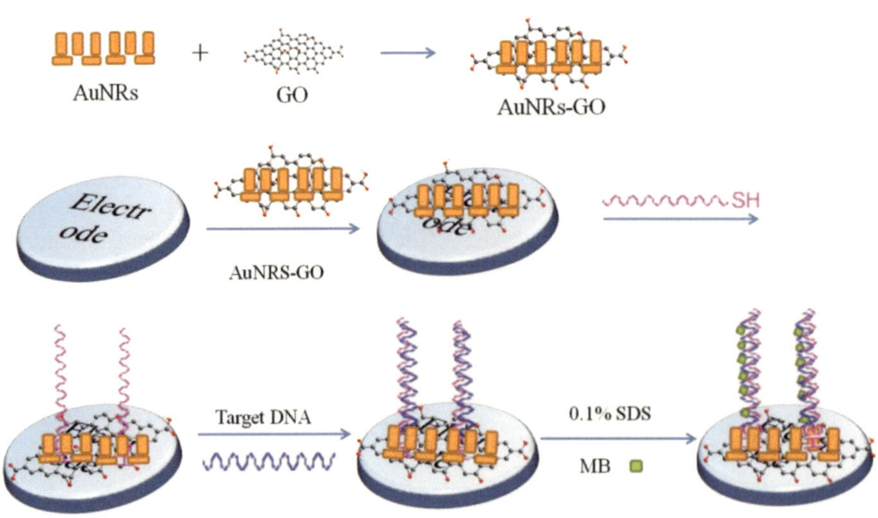

Figure 16. The scheme of DNA/AuNRs/GO biosensor [25].

http://dx.doi.org/10.5772/intechopen.78222

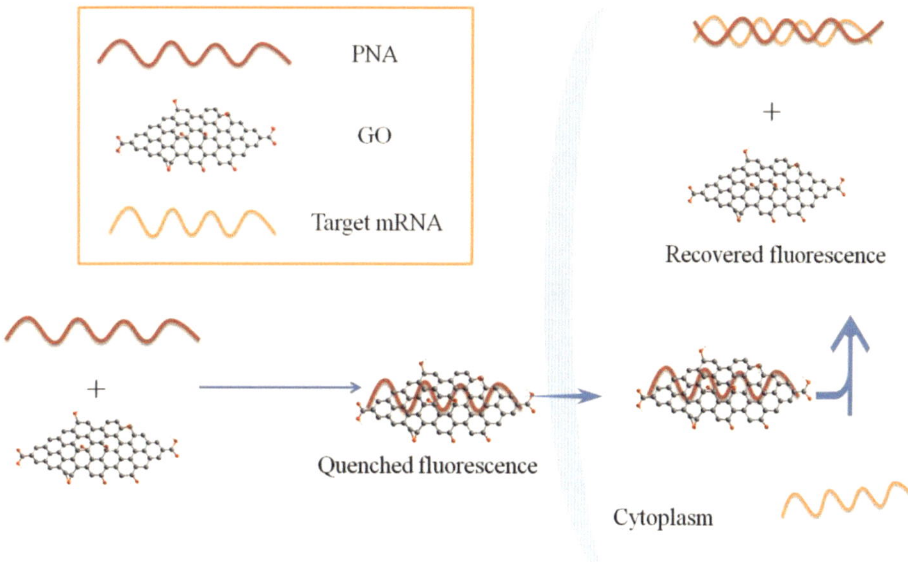

Figure 17. The scheme of PNA/NGO biosensor [26].

Zhang et al. [25] decorated gold nanorods (Au NRs) onto GO sheets and constructed a DNA biosensor (**Figure 16**). The AuNRs were prepared via a seed-mediated method and then composited with GO via electrostatic self-assembly. This biosensor exhibits significant selectivity and can distinguish complementary DNA in the presence of the 100-fold amount of single-base mismatched DNA.

Min et al. [26] designed a nano graphene oxide (NGO) based miRNA biosensor on evaluating target miRNA expression levels in living cells (**Figure 17**). The dye-labeled peptide nucleic acid (PNA) probes were binding onto the surface of NGO. In this biosensor, NGO and PNA acted as fluorescence quencher and probe, respectively. The miRNA expression levels can be evaluated by detecting the fluorescence quenching of the dye-labeled on PNA. The results showed that the biosensor exhibited a low detection limit (1 pM) and can detect the dynamic change in expression levers of the specific miRNA in stem cell differentiation [26].

4. Conclusion and outlook

Graphene oxide is one of many unique carbon materials, which displayed potential applications in the development of next-generation biosensors owing to its various physical and chemical properties. The functionalization of GO leads to the adsorption of various biomacromolecules, including enzymes such as glucose oxidase, horseradish peroxidase, laccase, and nucleic acids such as DNA and RNA for biosensing applications. The major prospect to be addressed in the future is the increasing demand for the engineering of biosensors based on GO that allow monitoring and detecting analytes with high selectivity and sensitivity at low cost. GO-based biosensors should also be fabricated as point-of-care devices for better in situ clinical diagnosis or as an in-situ sensing platform for environmental analysis.

Acknowledgements

High-Level Talent Start-Up Research Project of Foshan University (gg07016).

Conflict of interest

The authors declare no competing financial interest.

Notes/Thanks/Other declarations

The authors want to thank Bao-Yan Yang, Wei-Ming Gu, and Lin-Hao Lai from FoSU for their kindly help on editing presentation schemes and figures.

Acronyms and abbreviations

AP	1-aminopyrene
5′NH_2-ssDNA probe	5′-amine labeled single-stranded (ss) DNA probe
Cyt *c*	cytochrome *c*
DNA	deoxyribonucleic acid
ERGO	electrochemically reduced graphene oxide
Fc	ferrocene
FcCHO	ferrocenecarboxaldehyde
GCE	glassy carbon electrode
Au NRs	gold nanorods
GOD	glucose oxidase
GO	graphene oxide
GP	graphene platelet
G-AuNP	graphene-gold nanocomposite

http://dx.doi.org/10.5772/intechopen.78222

HRP	horseradish peroxidase
Lac	laccase
MWNTs	multiwall carbon nanotubes
ABEI	N-aminobutyl-N-ethylisoluminol
NGO	nano graphene oxide
RGO	reduced graphene oxide
RNA	ribonucleic acid
S/N	signal-to-noise ratio
SPCE	screen printed carbon electrode

Author details

Lingwen Zeng[1,2]*, Shilin Cao[1], Hang Yin[3], Jun Xiong[3] and Donghai Lin[2]

*Address all correspondence to: zeng6@yahoo.com

1 School of Food Science and Engineering, Foshan University, Foshan, China

2 Institute of Environment and Safety, Academy of Agricultural Sciences, Wuhan, China

3 School of Food Science and Engineering, South China University of Technology, Guangzhou, China

References

[1] Geim AK, Novoselov KS. The rise of graphene. Nature Materials. 2007;**6**:183-191

[2] Novoselov KS, Geim AK, Morozov SV, Jiang D, Zhang Y, et al. Electric field effect in atomically thin carbon films. Science. 2004;**306**:666-669

[3] Kuila T, Bose S, Khanra P, Mishra AK, Kim NH, et al. Recent advances in graphene-based biosensors. Biosensors & Bioelectronics. 2011;**26**:4637-4648

[4] Dreyer DR, Park S, Bielawski CW, Ruoff RS. The chemistry of graphene oxide. Chemical Society Reviews. 2010;**39**:228-240

[5] Wong CM, Wong KH, Chen XD. Glucose oxidase: Natural occurrence, function, properties and industrial applications. Applied Microbiology and Biotechnology. 2008; **78**:927-938

[6] Cao S-L, Xu H, Lai L-H, Gu W-M, Xu P, et al. Magnetic ZIF-8/cellulose/Fe_3O_4 nanocomposite: Preparation, characterization, and enzyme immobilization. Bioresources and Bioprocessing. 2017;**4**:56

[7] Kang X, Wang J, Wu H, Aksay IA, Liu J, et al. Glucose Oxidase-graphene-chitosan modified electrode for direct electrochemistry and glucose sensing. Biosensors & Bioelectronics. 2009;**25**:901-905

[8] Liu S, Tian J, Wang L, Luo Y, Lu W, et al. Self-assembled graphene platelet-glucose oxidase nanostructures for glucose biosensing. Biosensors & Bioelectronics. 2011;**26**:4491-4496

[9] Yang S, Lu Z, Luo S, Liu C, Tang Y. Direct electrodeposition of a biocomposite consisting of reduced graphene oxide, chitosan and glucose oxidase on a glassy carbon electrode for direct sensing of glucose. Microchimica Acta. 2013;**180**:127-135

[10] Qiu J-D, Huang J, Liang R-P. Nanocomposite film based on graphene oxide for high performance flexible glucose biosensor. Sensors and Actuators B-Chemical. 2011;**160**:287-294

[11] Palanisamy S, Vilian ATE, Chen S-M. Direct electrochemistry of glucose oxidase at reduced graphene oxide/zinc oxide composite modified electrode for glucose sensor. International Journal of Electrochemical Science. 2012;**7**:2153-2163

[12] Palanisamy S, Cheemalapati S, Chen S-M. Amperometric glucose biosensor based on glucose oxidase dispersed in multiwalled carbon nanotubes/graphene oxide hybrid biocomposite. Materials Science & Engineering C-Materials for Biological Applications. 2014;**34**:207-213

[13] Veitch NC. Horseradish peroxidase: A modern view of a classic enzyme. Phytochemistry. 2004;**65**:249-259

[14] Zhang Y, Lyu F, Ge J, Liu Z. Ink-jet printing an optimal multi-enzyme system. Chemical Communications. 2014;**50**:12919-12922

[15] Liu X, Han Z, Li F, Gao L, Liang G, et al. Highly chemiluminescent graphene oxide hybrids bifunctionalized by N-(aminobutyl)-N-(ethylisoluminol)/horseradish peroxidase and sensitive sensing of hydrogen peroxide. ACS Applied Materials & Interfaces. 2015;**7**:18283-18291

[16] Gu Z, Yang S, Li Z, Sun X, Wang G, et al. An ultrasensitive hydrogen peroxide biosensor based on electrocatalytic synergy of graphene-gold nanocomposite, CdTe-CdS core-shell quantum dots and gold nanoparticles. Analytica Chimica Acta. 2011;**701**:75-80

[17] Li C, Su Y, Lv X, Xia H, Wang Y. Electrochemical acetylene sensor based on Au/MWCNTs. Sensors and Actuators B-Chemical. 2010;**149**:427-431

[18] Zhang Q, Yang S, Zhang J, Zhang L, Kang P, et al. Fabrication of an electrochemical platform based on the self-assembly of graphene oxide-multiwall carbon nanotube nanocomposite and horseradish peroxidase: Direct electrochemistry and electrocatalysis. Nanotechnology. 2011;**22**:494010

[19] Zhang L, Cheng H, Zhang H-M, Qu L. Direct electrochemistry and electrocatalysis of horseradish peroxidase immobilized in graphene oxide-Nafion nanocomposite film. Electrochimica Acta. 2012;**65**:122-126

[20] Wan L, Song Y, Zhu H, Wang Y, Wang L. Electron transfer of co-immobilized cytochrome c and horseradish peroxidase in chitosan-graphene oxide modified electrode. International Journal of Electrochemical Science. 2011;**6**:4700-4713

[21] Liu H, Guo K, Lv J, Gao Y, Duan C, et al. A novel nitrite biosensor based on the direct electrochemistry of horseradish peroxidase immobilized on porous Co_3O_4 nanosheets and reduced graphene oxide composite modified electrode. Sensors and Actuators B-Chemical. 2017;**238**:249-256

[22] Palanisamy S, Unnikrishnan B, Chen S-M. An amperometric biosensor based on direct immobilization of horseradish peroxidase on electrochemically reduced graphene oxide modified screen printed carbon electrode. International Journal of Electrochemical Science. 2012;**7**:7935-7947

[23] Zhou X-H, Liu L-H, Bai X, Shi H-C. A reduced graphene oxide based biosensor for high-sensitive detection of phenols in water samples. Sensors and Actuators B-Chemical. 2013;**181**:661-667

[24] Singh A, Sinsinbar G, Choudhary M, Kumar V, Pasricha R, et al. Graphene oxide-chitosan nanocomposite based electrochemical DNA biosensor for detection of typhoid. Sensors and Actuators B: Chemical. 2013;**185**:675-684

[25] Han X, Fang X, Shi A, Wang J, Zhang Y. An electrochemical DNA biosensor based on gold nanorods decorated graphene oxide sheets for sensing platform. Analytical Biochemistry. 2013;**443**:117-123

[26] Ryoo S-R, Lee J, Yeo J, Na H-K, Kim Y-K, et al. Quantitative and multiplexed micro RNA sensing in living cells based on peptide nucleic acid and nano graphene oxide (PANGO). ACS Nano. 2013;**7**:5882-5891

Section 2

Graphene Oxide and Its Energy Conversion Applications

Chapter 6

Synthesis and Characterization of Reduced Graphene Oxide/Polyaniline/Au Nanoparticles Hybrid Material for Energy Applications

Carmina Menchaca-Campos, Elsa Pereyra-Laguna,
César García-Pérez, Miriam Flores-Domínguez,
Miguel A. García-Sánchez and
Jorge Uruchurtu-Chavarín

Additional information is available at the end of the chapter

http://dx.doi.org/10.5772/intechopen.77385

Abstract

In this work, synthesis and characterization of reduced graphene oxide/polyaniline/Au nanoparticles (GO/PANI/NpAu) as a hybrid capacitor are presented. Graphite oxide (GO) was synthesized by a modified Hummer's method. Polyaniline was synthesized by chemical polymerization, and Au nanoparticles (NpAu) were added afterward. Fabrication of the electrodes consisted on the hybrid materials being deposited on carbon cloth electrodes. The chemical and structural properties of the electrode were characterized by high-resolution scanning electron microscopy (HRSEM), Fourier transform infrared spectroscopy (FTIR), X-ray diffraction (X-R), and Raman spectroscopy; the results confirm the graphene reduction, the covalent functionalization, and formation of nanocomposites and also show the polyaniline grafted graphene. The performance and evaluation of the electrodes based on grapheme oxide (GO), polyaniline (PANI), GO-PANI, and GO/PANI/NpAu nanocomposites over carbon cloth, stainless steel, and copper have been obtained in 1 M H_2SO_4 solution, using electrochemical techniques namely: cyclic voltammetry (CV) and electrochemical impedance spectroscopy (EIS). They showed that GO/PANI/NpAu gave higher specific capacitance (SC) and energy values than PANI, and GO/PANI, in the order of 160 F/g. The present study introduces new hybrid material for energy applications, from the evaluation of their electrical contributions.

Keywords: graphene oxide, polyaniline, gold nanoparticles, electrochemical, redox

1. Introduction

With climate change and environmental concern, new energy sources have been created and various advanced energy storage systems. They simultaneously possess high energy and high power density as well as excellent recyclability, low cost, and friendly to the environment [1].

There are energy storage devices such as lithium-ion batteries often possessing high energy (200 Wh/Kg) but relatively low power density (1 KW/Kg), while traditional electrostatic capacitor has high power (40 KW/Kg) but low energy density (0.03 Wh/Kg) [2]. However, the electrochemical capacitor is a new type of high power and high energy density storage and deliverance device, therefore promising for feeding a variety of equipment operating with energy [3].

The capacitors can be divided into two classes based on charge-storage mechanism: (a) electrical double layer capacitors (EDLCs), where in electrode/electrolyte system, direction arrangement of the electron, or ion at the electrode/electrolyte interface forms electrical double layer [4] and (b) pseudo-capacitors, where the pseudo-capacitance arises from Faradaic reactions taking place at the electrode/electrolyte interface [5].

From the basic characteristics that determine the development of high-performance electrochemical capacitor (EC) electrodes, the most important are design, manufacture electrodes with suitable materials, architecture, and structure [2].

The typical electrode materials for EDLC are carbon materials (carbon nanotubes, graphene, activated carbon, etc.) due to their high-specific surface area and excellent conductivity [6], and the conductive polymers (polypyrrole, polyaniline, and polythiophene) are often used for pseudo-capacitors due to their high conductivity and large storage capacity [7]. However, the specific capacitance of carbon materials is commonly far less than that of conductive polymers, and the storage capacity of conductive polymers gradually decreases with the increase in the number of cycle (charge-discharge) [4].

In order to alleviate the inherent drawback of single materials, researchers have combined carbon materials and conductive polymers to obtain hybrid or composite materials with both high-specific capacitance and good cycle life called hybrid supercapacitors [8].

There are two methods for the preparation of hybrid capacitors: chemical methods in which an oxidizing agent is used and in-situ polymerization occurs, e.g., in situ chemical polymerization of graphene with polyaniline [9], forming the hybrid composite graphene/polyaniline (**Figure 1**). Another method is the electrochemical synthesis of conductive polymeric nanocomposites in which nanomaterials are dispersed in a monomer solution and formed by electropolymerization [10].

Being Au an excellent conductive material, it can be used to improve further the electrical properties of the GO/PANI hybrid material, incorporating it into the polymer matrix, as gold nanoparticles (NpAu), expecting the nanoscale dimension to potentiate the overall hybrid material properties.

With all the above ideas in mind, the next section shows some results about the synthesis of graphene oxide/polyaniline/Au nanoparticles hybrid material suitable to be used for energy applications.

http://dx.doi.org/10.5772/intechopen.77385

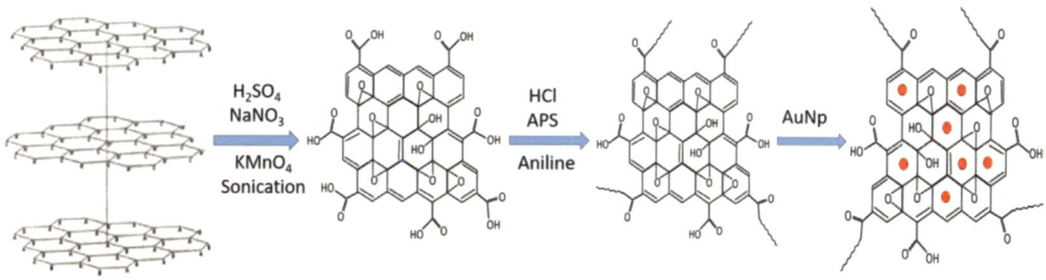

Figure 1. In situ polymerization of graphene oxide with polyaniline, forming the hybrid composite graphene oxide/polyaniline, and the incorporation of AuNp on the matrix GO/PANI.

2. Methodology development

The systems discussed were synthesized by the methodology, as shown in **Figure 1**, in which steps are described later and using the specified precursors and solvents.

2.1. Materials

All reactives were analytic degree: natural graphite powder, sulfuric acid 98% (H_2SO_4), hydrochloric acid (HCl), acetone, ethanol 98%, sodium nitrate ($NaNO_3$), and potassium permanganate ($KMnO_4$) from Meyer. Aniline monomer and gold (III) chloride trihydrate ($HAuCl_4 \cdot H_2O$) and polytetrafluoroethylene (PTFE) were purchased from Sigma Aldrich. The hydrogen peroxide (30% H_2O_2) was purchased from Reasol, and ammonium persulfate (APS) was from Golden Bell.

2.2. Synthesis of graphene oxide sheets

Graphite oxide (GO) was synthesized by a modified Hummer's method [11]. Graphite (10 g), $NaNO_3$ (5 g), and concentrated H_2SO_4 (230 ml) were mixed and stirred at 0°C in a 2000 ml reaction flask, which was immersed in an ice bath. Then, $KMnO_4$ (30 g) was added gradually over the stirring mixture, the temperature was controlled below 35°C, and the whole mixture was stirred for 2 h. After 30-min rest, the temperature of the mixture was raised to 98°C, and 460 ml of de-ionized water was slowly added to the suspension during 40 min.

After 30 min, the mixture was diluted by 1.4 l of de-ionized water and treated with (25 ml) H_2O_2 30% to reduce residual permanganate to soluble manganese ions until the gas evolution ceased. The resulting suspension was washed with HCl 1 M and de-ionized water until the filtrate became neutral and remaining impurities were removed. The product, graphite oxide, was exfoliated in an ultrasonic bath (2 h) to form graphene oxide (GO) sheets.

2.3. Synthesis of polyaniline nanofibers and in situ polymerization GO/PANI

Polyaniline was synthesized by chemical polymerization using ammonium persulfate (APS) as oxidant and HCl as doping agent. The aniline and APS mole ratio employed was 1:1, dissolved in 100 ml HCl 2 M separately, and put into an ice bath. The two solutions were mixed rapidly at 20°C temperature and put in ultrasonic bath for 3 h. The obtained green mixture was filtered and washed with ethanol and de-ionized water. The final product was put into a vacuum oven at 60°C for 4 h.

The polyaniline/graphene oxide (GO/PANI) composite was prepared by the same polymerization method with the presence of graphene oxide. In this case, the aniline was fixed at 40% wt. with 60% wt. graphene oxide in the acidic solution of HCl 2 M, according to the method reported and mentioned before [11].

2.4. Synthesis of Au nanoparticles (AuNp)

In a flask with 10 ml of 1 mM $HAuCl_4$ brought to boil with vigorous stirring, rapid addition of 1 ml of 38.8 mM sodium citrate to the vortex of the solution was added and resulted in a color change from yellow to burgundy. The heating at the same temperature and stirring was continued for an additional 15 min, and the resulting colloidal particles solution was stored at 4°C.

2.5. Electrodes and electrochemical measurement

To prepare the working electrode samples, GO, PANI or GO/PANI, and PTFE were mixed (90:10, w/w) and dispersed in ethanol. For the system GO/PANI, AuNp, and PTFE, they were mixed (72:20:8%), respectively, and were also dispersed by sonication in ethanol. Carbon cloth, stainless steel, and copper electrode (1 cm^2 area) were coated with the mixture and dried at room temperature for 12 h.

The electrochemical experiments were performed in a three electrode cell arrangement. A graphite rod was used as a counter electrode, and the potentials were measured with respect to a Ag/AgCl standard electrode (saturated with KCl). The electrochemical impedance measurements were carried out by applying an AC voltage of 10 mV amplitude in the 10 kHz–0.01 Hz frequency range. Cyclic voltammetry measurements were carried out in 1 M H_2SO_4 solution at different scan rates of 2–100 mV/s in a voltage range of −0.4 to 1.2 V. Electrochemical Gill AC Instruments analyzer ACM serial 1039 were used throughout the experiments.

3. Characterization of the hybrid materials

To realize the characterization of the synthesized hybrid materials, they were analyzed with the relevant methods and specified instruments.

3.1. Apparatus

Morphological characterization was carried out by using SEM Hitachi S-5500. The X-ray diffraction patterns were obtained from a Bruker D2 Phaser. The FTIR infrared analysis was measured with Platinum-ATR Alpha Bruker, and the ultraviolet visible spectra were obtained from Genesys 105 UV-Vis Thermo Scientific. Scanning electron microscope (SEM) was a LEO

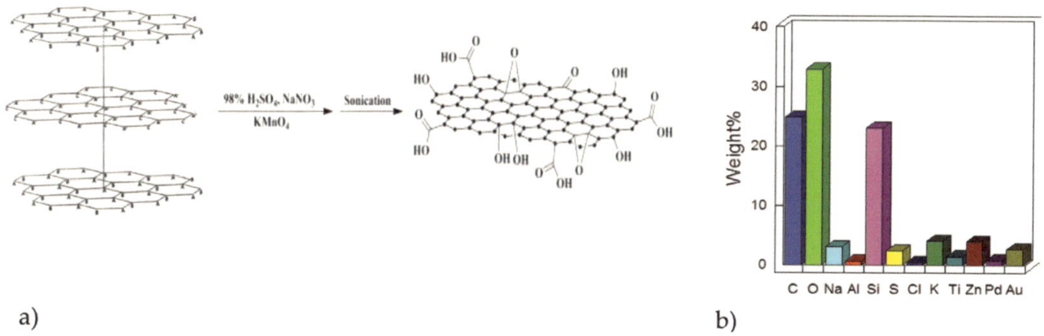

Figure 2. (a) GO sheets separation and (b) elemental analysis.

model operating at 15 kV, at 1, 5, 10, and 15 kX magnifications. Raman spectroscopy was a WITec alpha 300 AR, laser source: 532 nm (green), power: 15.6 mW, optical objective: 100×, integration time 5 s, eight accumulations.

3.2. Elemental analysis of the GO

As was mentioned in the methodology description, to obtain GO sheets, graphite was previously oxidized by a modified Hummer's method [11] consisting in the chemical oxidation of the structure through the use of concentrated sulfuric acid, potassium permanganate, and sodium nitrate. After oxidation, this was followed by ultrasonic bath, to break Van der Waals forces to separate the sheets and to obtain GO (**Figure 2a**). The incorporation of oxygen into the graphite crystalline network was corroborated, determining the carbon/oxygen ratio through SEM elemental X-ray analysis (**Figure 2b**).

3.3. FTIR spectroscopy

FTIR spectroscopy was used to elucidate the covalent grafting and to confirm the change in functional groups during each step. **Figure 3** represents the FTIR spectra of GO, PANI, GO/PANI, and GO/PANI/AuNp. The GO shows absorption bands at 3200 and 1734 cm^{-1}, which correspond to O—H, C=O in COOH [12]. It can be also observed that there are bands around 1605 and 1376, which are due to the intercalated water and deformation vibrations of C—O in C—OH and C—O—C functional groups [13].

For pure PANI prominent attributed absorption peaks are seen at 1630 and 1394 cm^{-1}, belonging to C=C stretching deformation of quinoid and C=N stretching of secondary aromatic amine, revealing the presence of emeraldine salt state in PANI [14]. The bands at 1184 and 805 cm^{-1} correspond to —C—N stretching vibration and out of plane bending vibrations of C—H in the benzene ring. Around 3281 cm^{-1}, it was observed an absorption band for N—H stretching of the amine group.

The FTIR spectrum of the GO/PANI composite was identical to that of PANI, which confirmed that the GO surface was wrapped by PANI [14]. There is no peak observed at 3200 and 1734 cm^{-1} (—OH and C=O vibrations, respectively), indicating the reduction of GO took

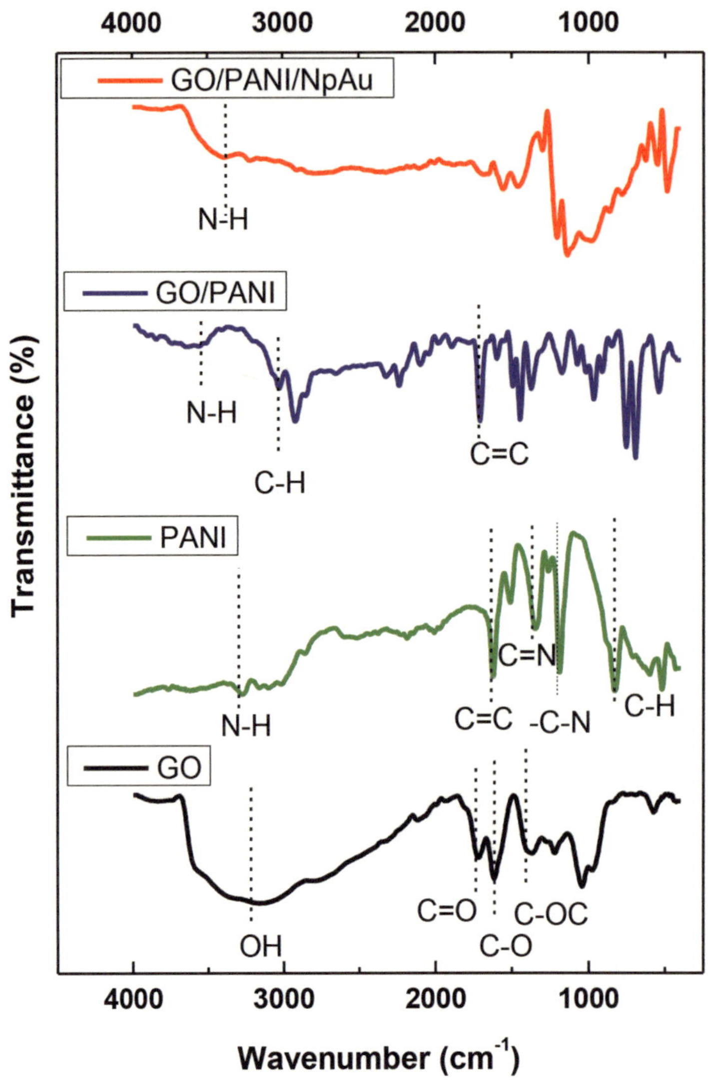

Figure 3. FT-IR spectrum of GO, PANI, and the composites GO/PANI and GO/PANI/AuNp.

place due to the polymerization of aniline, in addition to the confirmation that PANI has been covalently grafted onto the surface of the GO sheets. The FTIR bands were displaced, indicating a covalent bond and displacement due to the molecule arrangement. Functional groups responsible for stabilization of gold nanoparticles appeared in the band at 3300 cm^{-1} corresponding to N—H vibrations [15].

3.4. X-ray diffraction

Understating the morphological as well as structural changes of the products obtained, XRD studies on graphite, GO, PANI, GO/PANI, and GO/PANI/AuNp were analyzed.

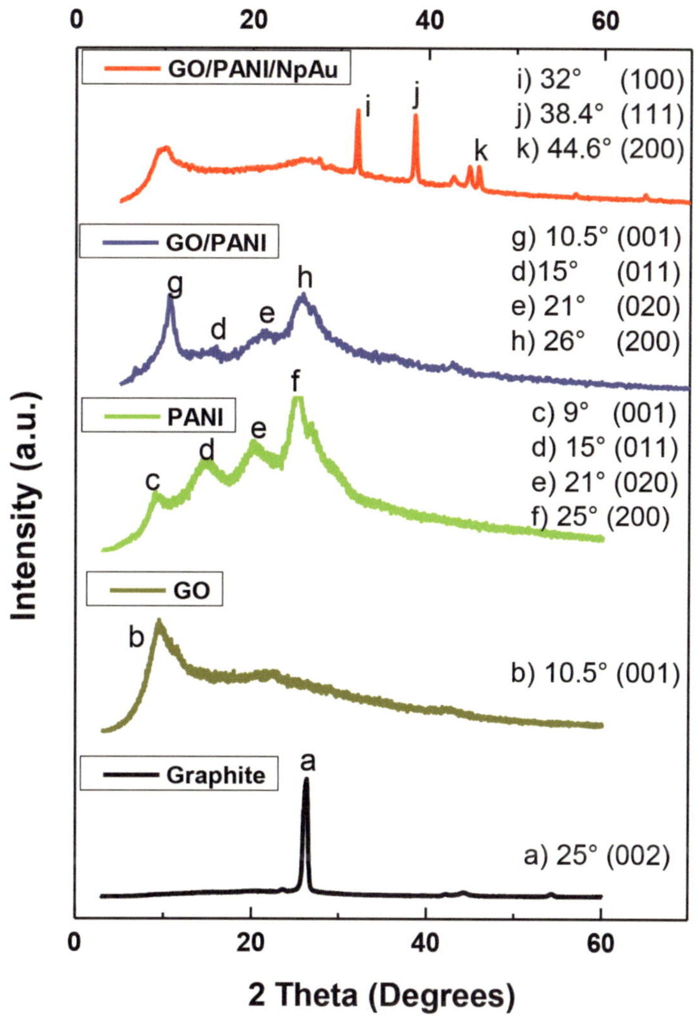

Figure 4. X-ray diffraction (XRD) of graphite, GO, PANI, GO/PANI, and GO/PANI/AuNp.

Figure 4 exhibits the XRD crystallographic pattern of graphite, and the basal reflection (002) peak at 2θ = 25° indicates a d-spacing of 0.35 nm based on Bragg's equation. After chemical oxidation, the peaks shifted to a lower angle reflection plane (001) at 2θ =10.5°, which indicates a d-spacing of 0.84 nm.

This widening of the d-spacing can be attributed to the intercalation of water molecules and generation of oxygenated functional such as epoxy, hydroxyl, and carboxyl groups between the inter-layering of the graphite sheets during severe oxidation [9], the small peak observed at 2θ = 42.5° associated with (100) plane of graphite, and indicates that small amounts of graphite phases are still present [10]. The peaks for emeraldine form PANI were exhibited at 2θ = 25, 21, 15, and 9° corresponding to (200), (020), (011), (001) reflections for PANI, respectively [16].

After the polymerization, it is observed the reduction of graphene oxide by the interaction with PANI demonstrated with the decrease of (001) reflection plane angle ($2\theta = 10.5°$). The XRD pattern of GO/PANI presents crystalline peaks similar to those of PANI. The peak around $2\theta = 26°$ is correlated to the interlayer space between the graphene sheets, which overlap with the diffractions from PANI [16]. The AuNp peaks were observed at 2θ = 32, 38.4, and 44.6°, corresponding to (100), (111), and (200) planes of the face centered cubic crystal, respectively.

3.5. Ultraviolet-vis

The AuNps were synthesized by Turkevich's method (1951) [17] in which reduction results were observed with the color change from pale yellow to burgundy. The color of AuNp is dependent upon the size and shape of the nanoparticles formed, which is correspondingly associated with the surface plasmon resonance due to collective oscillations of six electrons in the conduction band of AuNp, and this is the resonance frequency of the incident electromagnetic radiation [17, 18]. In **Figure 5a**, the burgundy particle dispersion attributed to circular shape particles with less than 40 nm in size can be observed, which agreed with circular shape with diameters 10-40 nm for AuNp, obtained by the citrate method [19].

The maximum absorption was observed at 520 nm (as seen in **Figure 5b**) corresponding to the plasmon collective oscillation of gold [20], indicating that the nanoparticles are evenly dispersed in the aqueous solvent. When the incident light wave frequency resonates with the electron coherent movement from the conduction band, it produces a strong absorption, which is the origin of the observed colloidal color [21]. For small metallic nanoparticles (less than 20 nm in diameter), the absorption spectrum only depends on the dipole oscillation, being the reason for the color change from the Au salt reaction (pale yellow) with the sodium citrate, forming gold nanoparticles (burgundy color), in which the resonant maximum absorption band of plasmon surface (520 nm) observed is AuNp characteristic nuclei response.

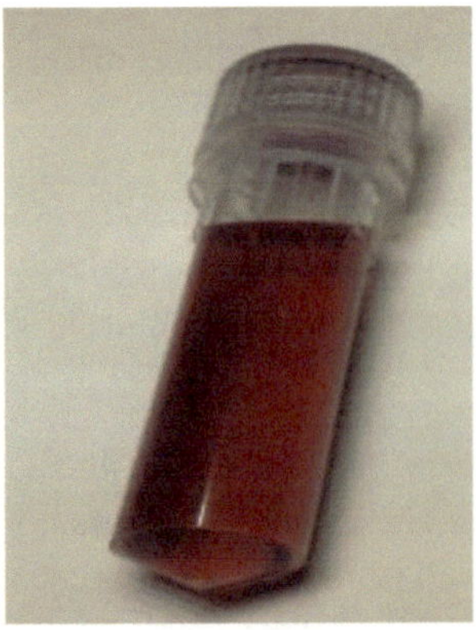

a)

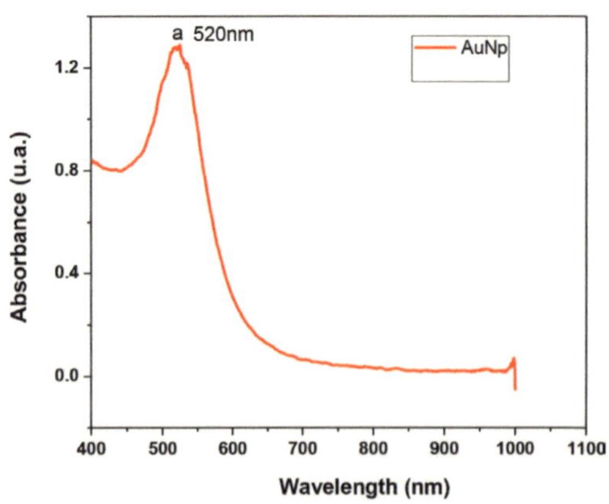

b)

Figure 5. UV-Vis of gold nanoparticles (a) nanoparticles dispersion and (b) the maximum absorption observed at 520 nm.

http://dx.doi.org/10.5772/intechopen.77385

Figure 6. HRSEM micrographs of (a–c) graphene oxide and (d) gold nanoparticles.

3.6. High-resolution scanning electron microscopy

High-resolution scanning electron microscopy (HRSEM) images were used to study size, shape, and dispersion about the synthesized nanomaterials. **Figure 6(a–c)** presents SEM micrographs at different magnifications showing views of the GO obtained by Hummer's modified method, graphene oxide sheets morphology are clearly seen, showing wrinkled and folded regions and the transparent property associated with them. Surface properties are associated with the shape and size of AuNp (**Figure 6d**) presenting sphere particles with an average size of 10–40 nm in diameter.

3.7. Raman spectroscopy

Raman spectroscopy is one of the most important techniques to detect and distinguish the graphene properties. This technique provides valuable information regarding the number of sheet layers, edges, and defects. This evaluation is very important because structural defects transform the graphene in such a way that intrinsic defects in the band structure modify its electric and magnetic properties.

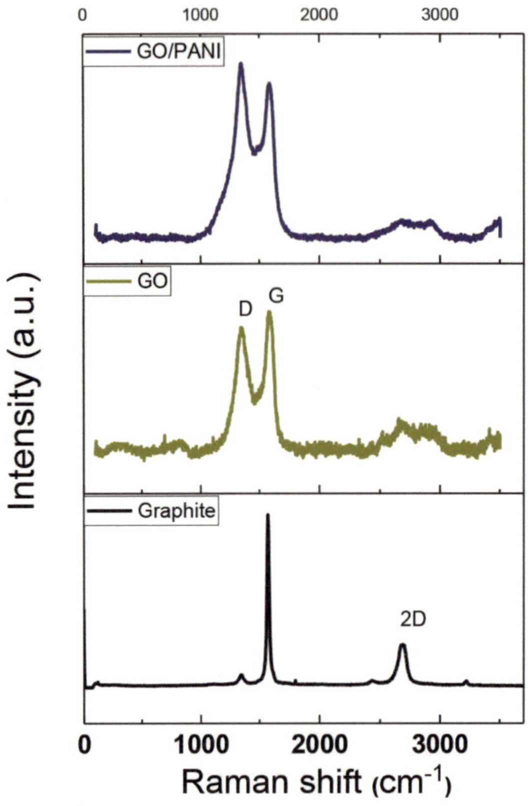

Figure 7. Raman spectra of graphite, GO, and GO/PANI composite.

Figure 7 shows the Raman spectra of graphite, GO, and GO/PANI composite. Graphite exhibits two peaks: a *D* band at 1363 cm^{-1} corresponding to defects or edge areas and a *G* band at 1577 cm^{-1} related to the vibration of sp^2-hybrided carbon. When GO is intercalated and oxidized, the band shifts to a higher wavenumber (1589 cm^{-1}) and widens as a result of a loss of interaction between the adjacent layers. The band of GO/PANI at 1358 cm^{-1} is more intense than that of GO due to the intercalation of oxygen-containing functional groups with covalent bonding in the GO layer [13, 14].

4. Electrochemical properties of the hybrid materials

4.1. Electrochemical capacitors

Electrochemical impedance spectroscopy (EIS) analysis was then carried out to investigate the electrochemical behavior of the materials synthesized. Hybrid coatings were tested first in carbon cloth acting as a substrate (blank). A typical Nyquist plot consisted of a semicircle in the high-frequency region and a linear part in the low-frequency area is formed [22]. Alternatively, the Bode plot presents the total impedance and phase angle as a function of frequency.

The diameter of the semicircle correlates with the interfacial charge-transfer resistance, usually representing the electrochemical reaction on the electrode (Faradaic resistance) [23]. All

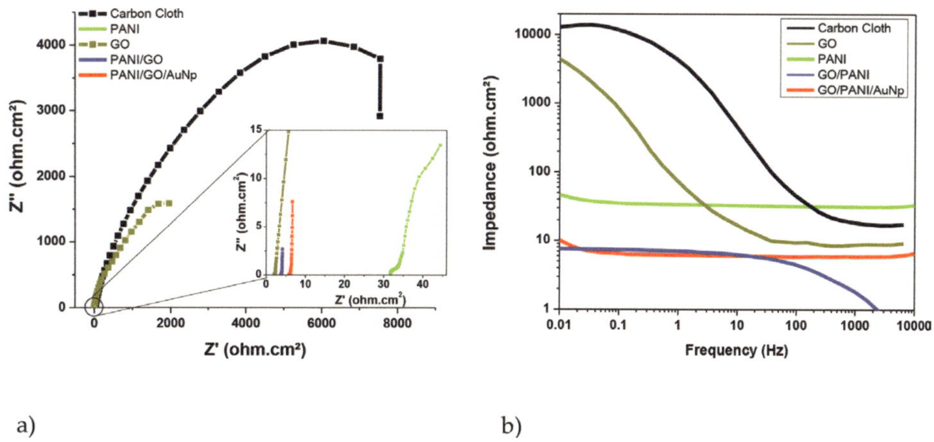

Figure 8. Impedance plots (a) Nyquist and (b) Bode for carbon cloth material support, and each synthesized material; GO, PANI, GO/PANI, GO/PANI/NpAu.

hybrid materials show very small diameter semicircle demonstrating the good electrical conductivity of the three-element composite, with a semi-straight line at mid to low frequencies, associated to capacitive behavior (mass transfer control) and lower charge-transfer resistance.

The Nyquist and Bode plots obtained are presented in **Figure 8a** and **b**, respectively, and for comparison purposes, the carbon cloth presents the highest overall or total impedance value above 1 kohm·cm^2 at the low frequency limit (0.01 Hz). The GO system shows a lower impedance value around 4500 kohms·cm^2, both showing an inverse frequency behavior at intermediate frequencies corresponding to double layer capacitance. For the PANI system, the impedance values diminished in all the frequency bandwidth considered around 45 ohms·cm^2, reflecting the conductive properties of the polymer material.

The GO/PANI/AuNp hybrid exhibits the semicircle smaller than GO/PANI and PANI, suggesting better conductivity and lower charge transfer resistance. A straight sloping line in the lower frequency represents the diffusion resistance (Warburg impedance, W), which reflects the diffusion or mass transfer of redox species in the electrolyte, and a steeper line usually indicates faster ion diffusion.

Based on these electrochemical analyses, the enhanced capacitive behavior was due to the synergistic effect between graphene oxide and PANI, besides the high conductivity of the AuNp. In addition, the small nanometer size can exhibit enhanced electrode/electrolyte interface areas, providing high electro-active regions and short diffusion lengths. This is also true for the GO/PANI and GO/PANI/AuNp hybrid materials showing a further decrease of around 10 ohms·cm^2, presenting low impedance behavior all around [22].

Bode impedance diagrams (**Figure 8**) demonstrate that the synthesized materials PANI, GO/PANI, and GO/PANI/AuNp present good conductivity properties with low impedance values. Also, the capacitive behavior was observed with the carbon cloth (blank) and the GO material. This analysis reveals that the good electrical conductivity and ion diffusion behavior resulted in the electrochemical performance of GO/PANI/AuNp of the three element hybrid material [23].

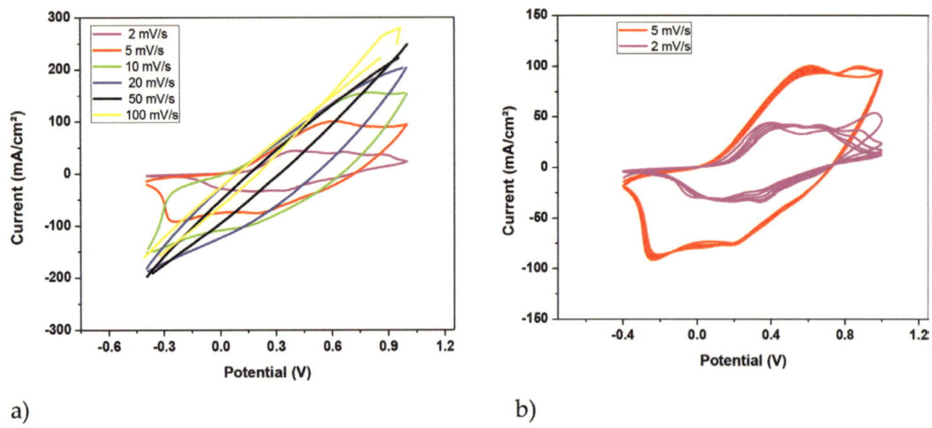

Figure 9. Cyclic voltammetry results of (a) carbon cloth different rates and (b) carbon cloth at 2 and 5 mV/s showing a trapezoidal form.

Another phenomenon that reinforces the theoretical explanation [24] is the reduction of the total impedance about two to three orders of magnitude with respect to the carbon cloth (blank) as can be seen in the Bode impedance diagram (**Figure 8b**), indicating the greatly increased ionic conductivity (high frequency) of the system.

Cyclic voltammetry is a tool used to evaluate the faradaic and nonfaradaic processes as well as the capacitive properties and mechanisms of reaction at the electrode interface (**Figure 9**). The electrochemical response of the carbon cloth, of current as a function of potential for different sweep rates, is shown (**Figure 9a**), obtaining the best rate at 5 mV/s, where a more trapezoidal shape was obtained (**Figure 9b**).

Figure 10a presents the cyclic voltammetry of carbon cloth coated with polyaniline (PANI) showing oxidation peaks for PANI around 0.5 and 0.96 V, while reduction peaks are observed at −0.2 and 0.2 V. The separation between peaks is more than 0.2 V; therefore, the reaction is quasi-reversible and attributed to transition of PANI, from emeraldine to the pernigraniline form [25]. The incorporation of PANI as a cover demonstrates the increased capacitance from the shape obtained.

In **Figure 10b**, the cyclic voltammogram for the different components applied to the carbon cloth supporting material (blank) in a (H_2SO_4) 1 M solution is presented. The graphs show when incorporating the PANI to the GO matrix the current increases about 63 mA/cm^2 as the maximum peak in the anodic current (ia), while an increase in 100 mA/cm^2 is observed as the maximum peak, for the whole hybrid system GO/PANI/AuNp. The subsequent incorporation of the different components to the hybrid system increases the total area of the trapezoid shape.

This involves two types of capacitive behavior contribution: from the electrochemical double layer capacitance (EDL) produced by GO and pseudo capacitive behavior from the incorporation of PANI. This suggestion is obtained from the two peaks observed in the voltammogram that indicated the existence of faradaic processes. This suggests a good synergism from the components of the matrix as well as good conductive properties [26].

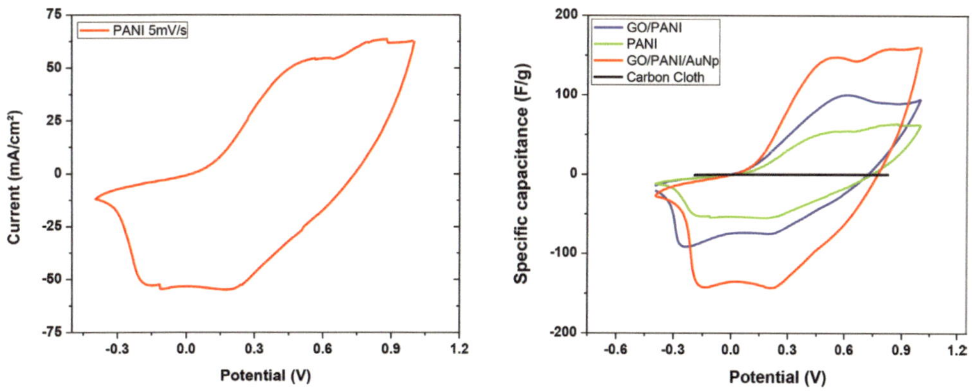

a) b)

Figure 10. Cyclic voltammetry results of (a) PANI showing redox peaks to transition of PANI, from leucoemeraldine/emeraldine and emeraldine/pernigraniline form and (b) carbon cloth, PANI, GO/PANI, and GO/PANI/AuNp.

The specific capacitance of the electrodes can be calculated according to the following equation (1) from CV curves:

$$C = \frac{I}{mV} \quad (1)$$

where I is the current, m is the mass of reactive material, and V is the potential scan rate.

The system presents potential to be used as capacitors with obtained values around 100–160 F/g. Au nanoparticles improve the capacitance behavior in GO/PANI/AuNp hybrid material for energy application.

4.2. Fuel cell electrodes

Another interesting possibility for the hybrid material synthesized may have promising applications as conductive material such as electrodes for fuel cells, in combination with metallic substrates. In the following examples, metal substrate includes copper and stainless steel [27].

For stainless steel substrate with the four synthesized hybrid component as coatings, immersed in the H_2SO_4 solution, the Nyquist and Bode impedance plots are presented in **Figure 11**. The two time constant behavior can be seen. The coatings response and dielectric properties are ascribed to higher frequencies (kHz), whereas at mid to lower frequencies (Hz–mHz), the response is associated with the oxide layer or bare metal substrate coating interface [28]. For the system tested, Nyquist plots present small diameter semicircle followed by a straight line with different slopes, corresponding to mass transfer diffusion process (**Figure 11a**). Bode plots (**Figure 11b**) present at the highest frequency: in the PANI coated stainless steel the highest impedance (70 ohms·cm2), followed by GO, GO/PANI, PANI and the lowest for GO/PANI/AuNp (10 ohms·cm^2).

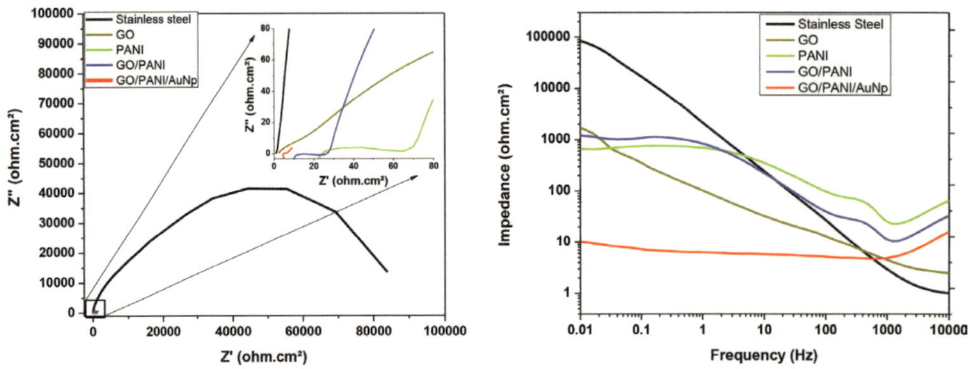

a) b)

Figure 11. Impedance for stainless steel and different coating components immersed in H_2SO_4 1 M solution (a) Nyquist and (b) Bode plots.

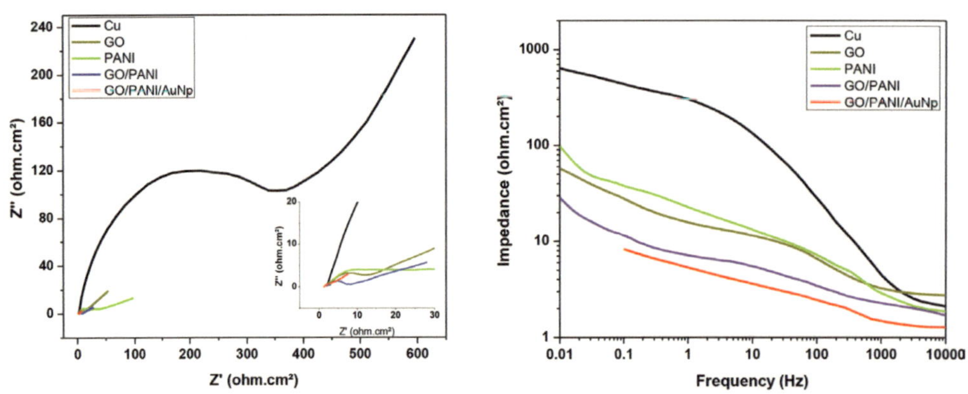

a) b)

Figure 12. Impedance for copper and different coating components immersed in H_2SO_4 1 M solution (a) Nyquist and (b) Bode plots.

In a similar way as before, the coating systems were applied, but this time in a copper metal substrate. A similar response was obtained (**Figure 12**), where the Nyquist plots show a semicircle with lower impedance resistance values as compared with stainless steel samples; followed by a straight line with different slopes associated to the coating protection and mass transport (**Figure 12a**) [29].

The high frequency impedance obtained was for the GO, followed by Cu, PANI, GO/PANI, and the lowest impedance value (15 ohms·cm^2) for the GO/PANI/AuNp coating sample. The highest impedance modulus (low frequency) seen is for the bare copper sample (650 ohms·cm^2) lowering its values for PANI, GO, and GO/PANI, and the smallest value registered is for the coating GO/PANI/AuNp system (**Figure 12b**). The mass transfer is reflected in the impedance modules obtained [30].

Both metal coating systems present good conductive properties for the hybrid synthesized coating system. Lower resistance, good ionic conductivity, and capacitive properties, which present GO/PANI/AuNp compound, make it attractive for different technological developments and applications in the energy sector.

5. Summary

It was possible to synthesize the hybrid material as evidenced by the characterization techniques. Electrochemical analysis demonstrates the capacitive redox behavior of PANI, but graphene, which is a material with large surface area (the property that increases the capacitance) when combined with the polyaniline conducting polymer, acts as a good material for electrical conduction rather than good capacitive behavior. The introduction of the AuNp in the network further increases the electric conduction. The trapezoidal shape of the voltammetric curves indicates maximum peaks with a capacitance of 160 F/g. This quantity indicates that the current is delivered very fast, which is not convenient for a capacitor. These results conclude that it is not a very good material for energy storage; however, it has promising applications as conductive material such as electrodes for fuel cells. Good electrochemical properties were obtained for metal electrodes coated with the hybrid systems, suggesting good energy electrode applications.

Acknowledgements

The authors wish to thank SEP-PROMEP for the support provided to both the Academic Body "Desarrollo y Análisis de Materiales Avanzados" (UAEMOR-CA-43) and the Academic Network "Diseño Nanoscópico y Textural de Materiales Avanzados." Finally, the authors thank CONACyT for the student grants received during this work.

Conflict of interest

The authors confirm that this content has no conflict of interest.

Author details

Carmina Menchaca-Campos[1], Elsa Pereyra-Laguna[1], César García-Pérez[1], Miriam Flores-Domínguez[1], Miguel A. García-Sánchez[2] and Jorge Uruchurtu-Chavarín[1]*

*Address all correspondence to: juch25@uaem.mx

1 Centro de Investigación en Ingeniería y Cciencias Aplicadas-(IICBA), Universidad Autónoma del Estado de Morelos, Cuernavaca, Morelos, Mexico

2 Departamento de Química, UAM-Iztapalapa, Mexico, DF, Mexico

References

[1] Fan W, Zhang C, Tjiu WW, Pramoda KP, He C, Liu T. Graphene-wrapped polyaniline hollow spheres as novel hybrid electrode materials for supercapacitor applications. ACS Applied Materials and Interfaces. 2013;**5**:3382-3391. DOI: 10.1021/am4003827

[2] Wu H, Zhang Y, Cheng L, Zheng L, Li YQ, Yuan W, Yuan X. Graphene based architectures for electrochemical capacitor. Energy Storage Materials. 2016;**5**:8-32. DOI: 10.1016/j.ensm 2016.05.003

[3] Sun H, She P. Xu K, Shang Y, Yin S, Liu Z. A self-standing nanocomposite foam of polyaniline reduced graphene oxide for flexible super-capacitor. Synthetic Metals. 2015;**209**:68-73. DOI: 10.1016/j.synthmet.2015.07.001

[4] Chen N, Ren Y, Kong P, Tan L, Feng H, Luo Y. In situ one-pot preparation of reduced graphene oxide/polyaniline composite for high performance electrochemical capacitor. Applied Surface Science. 2017;**392**:71-79. DOI: 10.1016/j.apsusc.2016.07.168

[5] Gupta S, SG P. Investigating graphene/conducting polymer hybrid layered composites as pseudocapacitors: Interplay of heterogeneous electron transfer, electric double layer and mechanical stability. Composites Part B Engineering. 2016;**105**:46-59. DOI: 10.1016/j.compositesb.2016.08.035

[6] Thirumal V, Pandurangan A, Jayavel R, Ilangovan R. Synthesis and characterization of boron doped graphene nanocheets for supercapacitor applications. Synthetic Metals. 2016;**220**:524-532. DOI: 10.1016/j.synthmet.2016.07.011Get

[7] West RM, Semancik S. Interpenetrating polyaniline-gold electrodes for SERS and electrochemical measurements. Applied Surface Science. 2016;**387**:260-267. DOI: 10.1016/j.apsusc.2016.06.073

[8] Liu Q, Nayfeh O, Nayfeh MH, Yao ST. Flexible supercapacitor sheets based on hybrid nanocomposite materials. Nano Energy. 2013;**2**:133-137. DOI: 10.1016/j.nanoen.2012.08.007

[9] Xiong P, Zhu J, Wang X. Recent advances on multi-component hybrid nanostructures for electrochemical capacitors. Journal of Power Sources. 2015;**294**:31-50. DOI: 10.1016/j.jpowsour.2015.06.062

[10] Shayeh JS, Ehsani A, Ganjali MR, Norouzi P, Jaleh JB. Conductive polymer/reduced graphene oxide/Au nano particles as efficient composite materials in electrochemical supercapacitor. Applied Surface Science. 2015;**353**:594-599. DOI: 10.1016/j.apsusc.2015.06.066

[11] Fernández PS. Modificación superficial de materiales de carbono: grafito y grafeno. Universidad de Oviedo; 2011. http://hdl.handle.net/10651/12792

[12] Xu Y, Liu Z, Zhang X, Wang Y, Tian J, Huang Y, Ma YY, Zhang X, Chen Y. A graphene hybrid material covalently functionalized with porphyrin: Synthesis and optical limiting property. Advanced Materials. 2009;**21**:1275-1279. DOI: 10.1002/adma.200801617

[13] Hu F, Li W, Zhang J, Meng W. Effect of gaphene oxide as a dopant on the electrochemical performance of graphene oxide/polyaniline composite. Journal of Materials Science and Technology. 2014;**30**:321-327. DOI: 10.1016/j.jmst.2013.10.009

[14] Kumar M, Singh K, Dhawan SK, Tharanikkarasu K, Chung JS, Kong BS, Kim EJ, Seung JK, Hur H. Synthesis and characterization of covalently-grafted graphene-polyaniline nanocomposites and its use in a supercapacitor. Chemical Engineering Journal. 2013;**231**:397-405. DOI: 10.1016/j.cej.2013.07.043

[15] Eisa WH, Zayed MF, Abdel-Moneam YK, Zeid AMA. Water-soluble gold/polyaniline core/shell nanocomposite: Synthesis and characterization. Synthetic Metals. 2014;**195**:23-28. DOI: 10.1016/j.synthmet.2014.05.012

[16] Zhao M, Wu X, Cai C. Polyaniline nanofibers: Synthesis, characterization and applications to direct electron transfer of glucose oxidase. Journal of Physical Chemistry. 2009;**113**:4987-4996. DOI: 10.1021/jp807621y

[17] Chen M, He Y, Liu X, Zhu J. Synthesis and optical properties of size-controlled gold nanoparticles. Powder Technology. 2017;**311**:25-33. DOI: 10.1016/j.powtec.2017.01.087

[18] Gharatape A, Salehi R. Recent progress in theranostic applications of hybrid gold nanoparticles. European Journal of Medicinal Chemistry. 2017;**138**:221-233. DOI: https://doi.org/10.1016/j.ejmech.2017.06.034"\t"_blank"\o "Persistent link using digital object identifier" 10.1016/j.ejmech.2017.06.034

[19] Grabar KC, Freeman RGW, Natan MJ, Khatherine C. Preparation and characterization of Au colloid monolayers. Analytical Chemistry. 1995;**67**:735-743. DOI: 0003-2700/95/0367-0735$9

[20] Hyperlink"https://www.sciencedirect.com/science/article/pii/S2405580817301103"\l "%21" Santhoshkumar Hyperlink "https://www.sciencedirect.com/science/article/pii/S2405580817301103"\l"%21"S.RajeshkumarHyperlink"https://www.sciencedirect.com/science/article/pii/S2405580817301103" \l "%21" Santhoshkumar J, Rajeshkumar S, Kumar SV. Phyto-assisted synthesis, characterization and applications of AuNPs–A review. Biochemistry and Biophysics Reports. 2017;**11**:46-57. DOI.org/10.1016/j.bbrep.2017.06.004 DOI: "https://doi.org/10.1016/j.bbrep.2017.06.004" \t "_blank" \o "Persistent link using digital object identifier" 10.1016/j.bbrep.2017.06.004

[21] Hyperlink"https://www.sciencedirect.com/science/article/pii/S0022231315004834" \l"%21"DasR,Hyperlink"https://www.sciencedirect.com/science/article/pii/S0022231315004834" \l "%21" Sarkar S, Hyperlink "https://www.sciencedirect.com/science/article/pii/S0022231315004834" \l "%21" Saha M, Hyperlink "https://www.sciencedirect.com/science/article/pii/S0022231315004834" \l "%21". Dey PC, Nath SS. Hyperlink "javascript:void(0)" Two peak luminescence from linoleic acid protected gold nanoparticles. Journal of Luminescence. 2015;**168**:325-329. DOI: Hyperlink "https://doi.org/10.1016/j.jlumin.2015.08.047" \t"_blank" \o"Persistent link using digital object identifier" 10.1016/j.jlumin.2015.08.047

[22] Hyperlink"https://www.sciencedirect.com/science/article/pii/S245191031630031X" \l "%21" Pajkossy T, Hyperlink"https://www.sciencedirect.com/science/article/pii/S245191031630031X" \l "%21" Jurczakowski R. Electrochemical impedance spectroscopy in interfacil studies. Current Opinion in Electrochemistry. 2017;**1**:53-58. DOI: Hyperlink "https://doi.org/10.1016/j.coelec.2017.01.006" \t "_blank" \o "Persistent link using digital object identifier" 10.1016/j.coelec.2017.01.006

[23] Hyperlink "https://www.sciencedirect.com/science/article/pii/S0379677915300096" \l "%21" Sun H, Hyperlink "https://www.sciencedirect.com/science/article/pii/S0379677915300096" \l "%21" She P, Hyperlink "https://www.sciencedirect.com/science/article/pii/S0379677915300096" \l "%21" Xu K, Hyperlink "https://www.sciencedirect.com/science/article/pii/S0379677915300096" \l "%21" Shang Y, Hyperlink "https://www.sciencedirect.com/science/article/pii/S0379677915300096" \l "%21" Yin S, Liu Z. A self-standing nanocomposite foam of polyaniline@reduced graphene oxide for flexible super-capacitor. Synthetic Metals. 2015;**209**:68-73. DOI: Hyperlink "https://doi.org/10.1016/j.synthmet.2015.07.001" \t "_blank" \o "Persistent link using digital object identifier" 10.1016/j.synthmet.2015.07.001

[24] Du W, Wang Z, Zhu Z, Hu S, Zhu X, Shi Y, Pang H, Qian X. Facile synthesis and superior electrochemical performances of CoNi/graphene nanocomposite suitable for supercapacitor electrodes. Hyperlink "https://doi.org/10.1039/2050-7496/2013" \o "Link to journal home page" Journal of materials chemistry A. 2014;**2**:9613-9619. DOI: "https://doi.org/10.1039/C4TA00414K" \t "_blank" \o "Link to landing page via DOI" 10.1039/C4TA00414K

[25] Bustos-Terrones V, Serratos-Álvarez IN, Córdoba-Herrera G, Escobar V, Osiris J, Uruchurtu-Chavarín J, Menchaca-Campos C, Preparación y caracterización del complejo polianilina-fluconazol, como pigmento en un recubrimiento anticorrosivo. Revista Tendencias en Docencia e Investigación en Química. 2016;**2**:QM111

[26] Philip D. Honey mediated green synthesis of gold nanoparticles. Spectrochimica Acta Part A. 2009;**73**:650-653. DOI: Hyperlink "https://doi.org/10.1016/j.saa.2009.03.007" \t "_blank" \o "Persistent link using digital object identifier" 10.1016/j.saa.2009.03.007

[27] García-Pérez C, Menchaca-Campos C, Garcia-Sanchez MA, Pereyra E, Rodriguez O, Uruchurtu J. Nylon/porphyrin/graphene oxide fiber ternary composite, synthesis and characterization. Open Journal of Composite Materials. 2017;**7**:146-165. DOI: 10.4236/ojcm.2017.73009

[28] Hernandez M. Hyperlink "https://www.scientific.net/author/Juan_Genesca_1" Genesca J, Hyperlink "https://www.scientific.net/author/Claudia_Ramos" Ramos C, Hyperlink "https://www.scientific.net/author/Emilio_Bucio" Bucio E, Hyperlink "https://www.scientific.net/author/Jos%C3%A9_Guadalupe_Ba%C3%B1uelos" Bañuelos JG, Hyperlink "https://www.scientific.net/author/Alba_Covelo" Covelo A. Corrosion resistance of AA2024-T3 coated with graphene/sol-gel films. Solid State Phenomena. 2015;**227**:115-118. DOI: "https://doi.org/10.4028/www.scientific.net/SSP.227.115" 10.4028/www.scientific.net/SSP.227.115

[29] Menchaca C, Castañeda I, Soto-Quintero A, Guardián R, Cruz R. García-Sánchez MA, Uruchurtu J. Characterization of a "smart" hybrid varnish electrospun nylon benzotriazole copper corrosion protection coating. International Journal of Corrosion. 2012;**2012**:925958-925967. DOI: 10.1155/2012/925958

[30] Menchaca-Campos C, García-Pérez C, Castañeda I, García-Sánchez MA, Guardián R, Jorge Uruchurtu J. Nylon/graphene oxide electrospun composite coating. International Journal of Polymer Science. 2013;**2013**:621627. DOI: 10.1155/2013/621618

Chapter 7

Defect Engineering in Reduced Graphene Oxide toward Advanced Energy Conversion

Guilherme Luís Cordeiro

Additional information is available at the end of the chapter

http://dx.doi.org/10.5772/intechopen.77386

Abstract

Defect engineering in reduced graphene oxide (rGO) for a smart design of fuel-cell supports has become an effective approach to improve the restricted two-dimensional (2D) mass and charge transfer and to boost the alcohol oxidation reaction. The present mini-review describes recent trends across prominent characteristics of tailored reduced graphene oxides, which include but are not restricted to, engineered three-dimensional (3D) nanostructures for better mass transport, tuned electron/hole conduction for easier electrical transport, and hybridized surfaces for high electrocatalytic activity. Special focus fixes upon the experimental progress on defect engineering, from three-dimensional structure assembly to surface metal complexation and heteroatom doping to size-controlled defect formation. Given their crucial impact on reduced graphene oxide properties, controlled methods for synthesis, and processing offer considerable promise toward next-generation carbon nanomaterials for electrocatalysis.

Keywords: defect engineering, reduced graphene oxide (rGO), electrocatalysis, fuel cells

1. Introduction

In response to an increasingly carbon-constrained world, the adoption of policies aimed at developing new technologies has emerged in the face of cleaner energy production. In this context, direct alcohol fuel cells (DAFCs) have been recognized as promising systems to provide continuous and low-carbon power supply. Basically, a DAFC operates by electrochemically oxidizing an alcohol, such as methanol or ethanol, at the anode, to produce protons (H^+ ions), and electrons. Protons are transferred to the cathode through the proton exchange membrane and react with OH^- ions, which are generated from the electrochemical reduction of oxygen, at the cathode, to produce water, heat, and electricity [1–4].

In spite of the attractiveness of these non-stop and low-carbon energy generation systems, important commercialization issues still need to be addressed. The kinetics of the alcohol oxidation reaction largely determines the overall efficiency of the fuel cell. In order to boost conversion efficiencies, highly active catalysts are required because of the low operating temperatures (60–120°C). Therefore, far, it is undisputed that platinum (Pt) provides the best correlation between energy adsorption and exchange current density [5]. With studies demonstrating the high instability of Pt catalysts [6–9] and the overall performance dependence on large Pt loadings [10–12], it has become imperative to design improved, durable, and highly efficient electrocatalysts.

Various attempts, such as the dispersion of Pt on high area conductive supports [13–19] and its combination with another metal [20–23] have been addressed for improving Pt utilization in fuel-cell reactions. Regarding the former approach, it is well-known that a suitable fuel-cell support provides a high surface-to-volume ratio of metal particles, which, in turn, maximizes the available area for electrochemical reactions [24]. In comparison to state-of-the-art C black, reduced graphene oxide (rGO) sheets have been demonstrated as an advanced electrocatalyst support for DAFCs due to the unique characteristics of the two-dimensional (2D) structure [24, 25]. The high theoretical surface area (2.630 $m^2 \cdot g^{-1}$ for a single layer) and ultra-large surface-to-volume ratio, when combined with the fast heterogeneous electron transfer (HET) rate, high specific capacitance (550 $F \cdot g^{-1}$), and intrinsic redox activity, make rGO an ideal platform for homogeneous dispersion of Pt nanoparticles and faster charge and mass transport properties [26–28].

In spite of the appealing properties noted above, restacking of the sheets due to the strong van der Waals interaction greatly reduces the accessible Pt surface area, resulting in low catalyst utilization, and transport pathway for reaction species. Material processing techniques, broadly defined as the approaches for tailoring physicochemical properties, have been extensively applied to control the interactions between rGO sheets and make them aggregation-resistant in both wet and solid state [29]. In this context, some solutions have been paving the way for further research and development on the assembly of two- or three-dimensional (3D) structures with desirable microstructural features for electrocatalysis. Positive progresses, such as the development of intercalation composites [30–34] and the usage of geometrical modification strategies [35–37], have greatly improved the utilization of supported Pt catalysts by increasing the density of exposed active sites.

Besides tailoring the physicochemical properties of the 2D structure, further advantageous characteristics for energy-conversion applications may be achieved by tuning the bandgap relative to the Dirac point in the C—C double bond network. Through electronic modulation of the support, high catalyst activity may be achieved by tuning the interaction between support and catalyst surfaces. In this sense, geometrically modified and/or heteroatom-doped rGO sheets, that is, can facilitate the property control of the Pt-support electronic effects. By enriching catalyst electronic structure due to catalyst/support synergism, novel characteristics, such as smaller catalyst particle size, increased catalyst particle dispersion, increased catalyst durability and stability, can effectively improve catalyst utilization [38–43].

Since the above observations reinforce the potential of intentional rGO modification as a strategy for boosting the 2D mass and charge transfer, defect engineering in rGO, which refers to the introduction of controlled defects in the material structure, is focused on this mini-review. Although this chapter makes no attempt to be exhaustive, the present contribution describes new breakthroughs on defect engineering in rGO that have recently been published since 2017, including recent advances and trends on state-of-the-art synthesis and utilization of engineered rGO sheets as fuel-cell support materials for the methanol and ethanol oxidation reactions. Future perspectives for further development are also proposed.

2. Overview of defect engineering in rGO for energy conversion

Defect engineering in 2D semiconductor technology refers to the introduction of controlled defects at the atomic level, such as heteroatoms and size-controlled vacancies, for the modification of the two-dimensional structure and properties. In spite of these, other two strategies, a 3D structure assembly approach and a surface metal complexation methodology, have been included as part of a broadened view of defect engineering in reduced graphene oxide, as summarized in **Figure 1**. To produce tailor-made support materials with desirable characteristics for fuel-cell catalysis, the usage and/or combination of defect-induced procedures is proposed toward the development of the next-generation rGO support materials.

In the first subsection of this mini-review, recent progresses on the synthesis and electrocatalysis of 3D engineered rGO-based platinum catalysts toward methanol oxidation reaction (MOR) and ethanol oxidation reaction (EOR) are discussed. Then, advances in heteroatom doping for designing highly conductive three-dimensional rGO-based platinum catalysts and the impacts on electrocatalysis are presented. In the last subsection, research directions on surface metal complexation and size-controlled defect formation through metal-organic frameworks (MOFs) are proposed as future perspectives of further development.

2.1. Current trends in 3D structure assembly for enhanced mass transfer

Crumpling the sheets into porous frameworks has been highlighted as an attractive methodology for enabling the interaction among nanoparticles and reactants. Kwok et al. [44] produced a high-quality platinum-decorated rGO aerogel with the aid of a solvothermal method. Their observations indicated that the rGO aerogel porous framework can be optimized by simply changing the GO concentration input for the gelation process. Tests on the supported ultrafine platinum nanoparticles (sizes ranging from 1.5 to 3 nm) showed that the electrochemically active surface area (ECSA) increased by about 8.92 times in comparison to a benchmark Pt/C catalyst, resulting in a 358% increment in specific power for a methanol-fed fuel cell.

In order to enhance catalyst utilization, Radhakrishnan et al. [45] fabricated a three-dimensional assembly of platinum nanostructures with dominant (100) plane on rGO by a co-electrodeposition method. They found that the morphology, active site, and the electrochemical activity of the catalyst were highly dependent on the number of electrochemical cycling used for the deposition. Their nanocomposite showed a high mass activity toward MOR, which

Figure 1. General schematic roadmap of defect engineering strategies in reduced graphene oxide support materials for achieving advanced energy conversion.

was attributed to the strong metal-support interactions. At the atomic level, the single-step approach for growing morphology-controlled nanostructures incurred in geometric and an electronic changes of platinum surface for enhanced mass/charge transfer.

Abundant mass transfer channels were recently introduced by a geometric change of rGO surface, as reported by Qiu et al. [46]. Following the development of a sacrificial template method, platinum nanoparticles on 3D reduced graphene oxide hollow nanospheres were synthesized. The enhanced activity for the methanol oxidation reaction was provided by the three-dimensional microporous structure, which facilitated the exposure of the active sites thereby promoting the ion and mass transfer processes. In addition, the improved activity of the electrocatalyst was ascribed to the electronic effects of platinum when alloyed with transition metals (iron, cobalt, nickel) for achieving low platinum loading in fuel cells.

Further research and development on material processing approaches may be illustrated by the recent work of Wang et al. [47]. In their contribution, intercalation was extended to a polyaniline (PANI) functionalization method. Deposited PANI effectively prevented rGO sheets from restacking during the preparation of the electrodes while a three-dimensional structure could also be produced during the process. Synergistic effect of the porous framework was associated to faster mass and electron transfer, resulting in superior methanol electrooxidation. The advantage of the novel synthesis procedure is that the conductivity of rGO could be preserved with the increase in porosity due to the conductive polymeric network created by intercalated PANI. In related work in terms of covalent functionalization, Pinithchaisakula et al. [34] also demonstrated that polydopamine PDA-functionalized reduced graphene oxide improved the MOR activity. Very recently, Waenkaew et al. [48] also pointed the effect of PDA on the electrocatalytic activity of Pt on rGO toward the oxidation of both methanol and ethanol.

As briefly pointed out, a three-dimensional hierarchical porous structure enables rGO sheets to efficiently improve the MOR activity due to the facilitated mass transfer arising from the 3D morphology. Revisited works indicate the superior advantage of an architectured reduced graphene oxide structure over conventional carbon black support materials. In light of recent developments [44, 47], further porous electrode optimization is highly recommended for maintaining the electrical conductivity of the rGO with the increase in porosity and decrease in density. By combining the heteroatom doping strategy, that is, to boost electrode conductivity, the porous materials will pave a way for fuel-cell development in the future.

2.2. Current trends in heteroatom doping for improved charge transfer

Efficiency increase in fuel cells has put forward a new prospect for the rational design of heteroatom-doped carbon nanomaterials for advanced energy conversion. This, together with the high-cost and scarcity of platinum has driven an intensive research effort for the development of metal-free cathode catalysts. Indeed, heteroatom-doped reduced graphene oxide has led to a huge amount of literature on metal-free catalysts for oxygen electrocatalysis [49, 50]. Nevertheless, doped carbon nanomaterials could largely impact anode technology, guiding advances on supported catalysts with low platinum content.

Since current graphene derivatives suffer from low intrinsic conduction as fuel-cell catalyst supports and current carriers, heteroatom doping, the process in which some carbon atoms are replaced by heteroatoms, could have a positive impact on the development of highly conductive support materials. By considering the most recent works on this hot topic, current trends indicate that heteroatom doping is capable of simultaneously providing proper nanoparticle dispersion and size-controlled active sites, and along with tuning catalyst d-band for achieving superior catalyst electrochemical activity [51–53].

Among various heteroatom doping strategies, the incorporation of nitrogen (N) has been the most studied because quaternary N atoms (substitutional N-doping), as depicted in **Figure 2**, can introduce high positive charge distribution in the nearby C atoms due to the high electron-withdrawing ability of N [50]. Furthermore, substitutional defects in the two-dimensional network can act as electron donors, providing n-type conductivity [54]. As a consequence, the former feature could lead to defective sites for efficient attachment of catalyst nanoparticles, whereas the latter could enhance an electron transport from support to the attached platinum nanoparticles, thereby increasing catalyst tolerance to poisoning.

Further development on the heteroatom doping approach may be followed with the work of Kanninen et al. [51]. In their methodology, reduced graphene oxide was co-doped with nitrogen

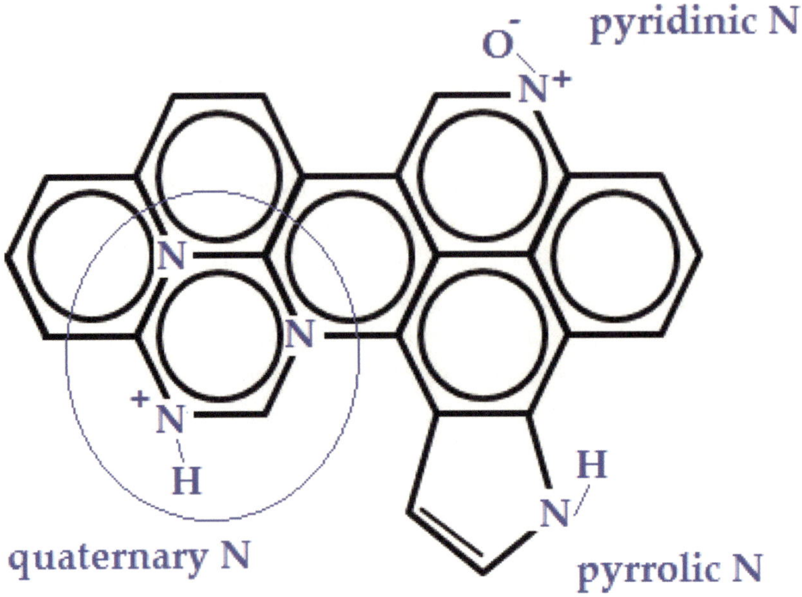

Figure 2. Schematic diagram of nitrogen dopant configuration in the 2D carbon—carbon network.

and sulphur atoms by a thiol-ene click chemistry approach. Co-doping changed platinum-support interaction, leading to a clear improvement on catalyst activity toward EOR in comparison to platinum supported on an undoped rGO. In addition, the induced defect sites generated from the co-doping strategy promoted the stabilization of the supported nanoparticles, resulting in enhanced catalyst durability. The latter was confirmed after a potential cycling test, revealing a minor loss in ECSA. After 500 potential cycles, the order of ECSA decrease was: Pt/C ($-12.3\ m^2 \cdot g^{-1}$) > Pt/rGO ($-3.8\ m^2 \cdot g^{-1}$) > Pt/rGO/ double-walled carbon nanotube (DWCNT) ($-2.7\ m^2 \cdot g^{-1}$) > Pt/NS-rGO-DWCNT ($-2.0\ m^2 \cdot g^{-1}$). Moreover, mixing double-walled carbon nanotubes (DWCNT) could prevent the restacking of the rGO sheets, leading to faster mass transfer.

Rethinasabapathy et al. [52] strategically tuned the MOR activity of supported platinum nanoparticles by designing an interconnected porous N-doped rGO through a hydrothermal method with urea as N source. They found that the incorporation of nitrogen atoms broke the 2D lattice symmetry, creating induced-defect sites for an efficient attachment of metal nanoparticles. Electrochemical characterization revealed a significant higher catalytic activity toward MOR in comparison to a nitrogen-free rGO support, which was attributed to the smaller particle size, narrower size distribution, and better dispersion of nanoparticles. The synergetic effect of N doping also contributed to an improvement of electronic conductivity, facilitating charge transfer from platinum-support interface to the strongly adsorbed MOR intermediates.

Inspired by the advantage of a three-dimensional assembly-assisted method, Zhao et al. [53] prepared a three-dimensional N-doped rGO through a feasible and simple one-pot hydrothermal process. By using melamine-cyanurate as both N source and pore forming agent, they reported the formation of an interconnected micrometer-scaled porous structure that could

http://dx.doi.org/10.5772/intechopen.77386

Catalyst	Synthesis conditions	Reaction	Performance parameter		Refs.
			ECSA ($m^2 \cdot g^{-1}$)	I* ($mA \cdot mg^{-1}$)	
Recent electrode materials tailored by the 3D structure assembly approach					
Pt/GO	Pt-precursor in 2 $mg \cdot mL^{-1}$ GO autoclaved @ 180°C/12 h	MOR	188 (~9× more than Pt/C)	–	[44]
rGO-Pt1	Thermal exfoliation of GO @ 180°C/3 h	MOR	>6 K (~3× more than Pt/C)	~2.5 K (~2.5× more than Pt/C)	[45]
Pt/3D-GNs (PANI)	Pt-precursor and Ani in 2 $mg \cdot mL^{-1}$ GO autoclaved @ 180°C/24 h	MOR	48 (~2.4× more than Pt/C)	~550 (~3.5× more than Pt/C)	[47]
Recent electrode materials tailored by the heteroatom doping approach					
Pt/NS-rGO/DWCNT	NS-functionalized GO by a thiol-ene click reaction @ 70°C/12 h	EOR	Stability test: 500 potential cycles. ECSA loss: Pt/C (−12.3 $m^2 \cdot g^{-1}$) and Pt/NS-rGO-DWCNT (−2.0 $m^2 \cdot g^{-1}$).		[51]
PtRuFe/NG	GO/urea (1:300 m/m) autoclaved @ 180°C/12 h	MOR	96 (~2× more than PtRuFe/rGO)	–	[52]
Pt/NGA	Melamine in 2 $mg \cdot mL^{-1}$ GO autoclaved @ 180°C/6 h followed by annealing @ 1000°C/1 h under Ar	MOR	60.6 (1.16× more than Pt/C)	~20 (~3× more than Pt/C)	[53]

*Current value obtained after chronoamperometry.

Table 1. Overview of the performance of some highly active electrode materials for alcohol electrocatalysis.

expose abundantly accessible catalyst sites, which led to an ECSA increased by 1.16 times in comparison to a benchmark Pt/C catalyst. As a result, platinum on rGO could enhance the MOR activity by approximately three times when compared to platinum on carbon black. Their recent results also revealed the advantage of N-doping, proper control of the platinum-support electronic effects, increasing catalyst particle dispersion, and decreasing catalyst particle size and holding a high prospect for application as an anode catalyst toward MOR.

Briefly, current methodologies for synthesizing defect-induced 3D C nanosheets were aimed on the design of high-performance rGO fuel-cell support materials and current carriers. Also, noteworthy is the influence of the electronic interaction between support and catalyst in electrocatalysis. Based on these, non-metal heteroatom doping in architectured or expanded stacks of reduced graphene oxide is proposed herein as an alternative for maintaining the electrical conductivity of the porous electrode. In addition, future trends toward superior electrocatalysis should extend the modulation of the electronic properties of both support and supported catalysts to incorporation of different non-metal heteroatoms, such as boron (B), fluorine (F), and sulphur (S), into the carbon—carbon 2D network. In **Table 1**, a summary of the performance of some highly active electrode materials is presented.

2.3. Future perspectives in defect engineering for superior electrocatalysis

Immense research work has been done in the field of materials science toward the study of MOFs for highly efficient oxygen electrocatalysis [55]. Basically, MOFs are a class of composite materials with a unique and highly ordered nanoporous (pore sizes smaller than 100 nm) structure, which are designed from a self-assembly of inorganic metals and organic linkers through co-ordination bonds. From an electrochemical aspect, co-ordinated metals could contribute to an electronic effect as a consequence of the changes in platinum electronic structure, directly influencing catalyst activity. Thanks to abundant dual-metal active sites, superior electrocatalysts with ultra-low platinum content could be developed for alcohol electrocatalysis further improving specific activity beyond the conventional platinum alloy approach.

Moreover, surface modification of rGO-supported platinum could lead to improved three-dimensional nano- (or micro-) structures, resulting in synergistic effects for achieving faster mass and charge transport properties. In this context, proper synthetic procedures based on the controlled carbonization of the organic linkers (primarily containing heteroatoms) might serve as novel rational design strategies for the development of highly porous heteroatom-doped supported catalysts. Also, noteworthy is that such a synthetic method, especially for large-scale production, should meet the requirements for commercialization, which include low cost, environmental sustainability and high reproducibility.

Last but not least, another contribution of the metal-organic framework process could be the *in situ* generation of metal oxide nanoparticles and their concomitant dispersion on rGO upon carbonization of the metal-organic complex. Thus, the deposited metal oxide nanoparticles could act as seeds for vacancy generation by etching the carbon—carbon network along the rGO-metal oxide interface. Size-controlled vacancies in the matrix of reduced graphene oxide are not only predicted to break the two-dimensional lattice symmetry, thereby tuning conduction mechanisms, but also act as trapping sites for heteroatom doping [56]. Therefore, in addition, heteroatom-doped porous and open structures on the basal planes together with the metal oxide nanoparticles might serve as novel active sites with high bifunctional activity. Precisely, a bifunctional effect has been associated to the presence of sites that aid in the dissociation of water to form surface hydroxides, which can readily oxidize strongly adsorbed reaction intermediates. Indeed, development of heteroatom-doped size-controlled vacancies could positively contribute toward the improvement of platinum activity in alcohol oxidation reactions.

3. Conclusions

As summarized in this mini-review, an ideal fuel-cell electrode should be porous, and possess high conductivity, accessible electrochemical surface sites, and improved charge and mass transfer pathways. Defect engineering, which involves manipulating the type, concentration, or spatial distribution of heteroatoms and size-controlled vacancies within a solid, along with materials processing approaches, such as three-dimensional structure assembly and surface metal complexation methodologies, has demonstrated its potential to tackle the challenges triggered by energy conversion concerns in direct alcohol fuel cells. With continuous progress on the knowledge gained from the engineered nanosheets, a transition from bench-scale nanotechnology to pilot plant manufactures and, eventually, commercial production is likely to be configured.

Acknowledgements

The author acknowledges the Brazilian National Council for Scientific and Technological Development (CNPq) for research funding.

Conflict of interest

The author declares no conflict of interest.

Abbreviations

B	boron
C	carbon
DAFC	direct alcohol fuel cell
DWCNT	double-walled carbon nanotube
ECSA	electrochemically active surface area
EOR	ethanol oxidation reaction
F	fluorine
H	hydrogen
HET	heterogeneous electron transfer
MOF	metal-organic framework
MOR	methanol oxidation reaction
N	nitrogen
O	oxygen
PANI	polyaniline
PDA	polydopamine
Pt	platinum
rGO	reduced graphene oxide
S	sulfur

Author details

Guilherme Luís Cordeiro

Address all correspondence to: gcordeiro@usp.br

Materials Science and Technology Center, Energy and Nuclear Research Institute (IPEN-CNEN/SP), São Paulo, Brazil

References

[1] Kordesch KV, Simader GR. Environmental impact of fuel cell technology. Chemical Reviews. 1995;**95**:191-207. DOI: 10.1021/cr00033a007

[2] Fuller TF, Perry ML. A historical perspective of fuel-cell technology in the 20th century. Journal of the Electrochemical Society. 2002;**149**:S59-S67. DOI: 10.1149/1.1488651

[3] Sharaf OZ, Orhan MF. An overview of fuel cell technology: Fundamentals and applications. Renewable and Sustainable Energy Reviews. 2014;**32**:810-853. DOI: 10.1016/j.rser.2014.01.012

[4] Wang Y, Chen KS, Mishler J, Cho SC, Adroher XC. A review of polymer electrolyte membrane fuel cells: Technology, applications, and needs on fundamental research. Applied Energy. 2011;**88**:981-1007. DOI: 10.1016/j.apenergy.2010.09.030

[5] Holton OT, Stevenson JW. The role of platinum in proton exchange membrane fuel cells. Platinum Metals Review. 2013;**57**:259-271. DOI: 10.1595/147106713x671222

[6] Shao-Horn Y, Sheng WC, Chen S, Ferreira PJ, Holby EF. Morgan. Instability of supported platinum nanoparticles in low-temperature fuel cells. Topics in Catalysis. 2007;**46**:285-305. DOI: 10.1007/s11244-007-9000-0

[7] Galeano C, Meier JC, Peinecke V, Bongard H, Katsounaros I, Topalov AA, Lu A, Mayrhofer KJJ, Schüth F. Toward highly stable electrocatalysts via nanoparticle pore confinement. Journal of the American Chemical Society. 2012;**134**:20457-20465. DOI: 10.1021/ja308570c

[8] Marcu A, Toth G, Kundu S, Colmenares LC, Behm RJ. Ex situ testing method to characterize cathode catalysts degradation under simulated start-up/shut-down conditions–A contribution to polymer electrolyte membrane fuel cell benchmarking. Journal of Power Sources. 2012;**215**:266-273. DOI: 10.1016/j.jpowsour.2012.05.010

[9] Hashimasa Y, Numata T. Comparison of test results on load cycle durability of polymer electrolyte fuel cell cathode catalysts. International Journal of Hydrogen Energy. 2015;**40**:11543-11549. DOI: 10.1016/j.ijhydene.2015.04.031

[10] Zhou W, Zhou Z, Song S, Li W, Sun G, Tsiakaras P, Xin Q. Pt based anode catalysts for direct ethanol fuel cells. Applied Catalysis B: Environmental. 2003;**46**:273-285. DOI: 10.1016/S0926-3373(03)00218-2

[11] Rousseau S, Coutanceau C, Lamy C, Léger J-M. Direct ethanol fuel cell (DEFC): Electrical performances and reaction products distribution under operating conditions with different platinum-based anodes. Journal of Power Sources. 2006;**158**:18-24. DOI: 10.1016/j.jpowsour.2005.08.027

[12] Glass DE, Olah GA, Prakash GKS. Effect of the thickness of the anode electrode catalyst layers on the performance in direct methanol fuel cells. Journal of Power Sources. 2017;**352**:165-173. DOI: 10.1016/j.jpowsour.2017.03.106

[13] Wang Y, Zou S, Cai W. Recent advances on electro-oxidation of ethanol on Pt- and Pd-based catalysts: From reaction mechanisms to catalytic materials. Catalysts. 2015;**5**:1507-1534. DOI: 10.3390/catal5031507

[14] Joo SH, Choi SJ, Oh I, Kwak J, Liu Z, Terasaki O, Ryoo R. Ordered nanoporous arrays of carbon supporting high dispersions of platinum nanoparticles. Nature. 2001;**412**:169-172. DOI: 10.1038/35084046

[15] Yoshitake T, Shimakawa Y, Kuroshima S, Kimura H, Ichihashi T, Kubo Y, Kasuya D, Takahashi K, Kokai F, Yudasaka M, Iijima S. Preparation of fine platinum catalyst supported on single-wall carbon nanohorns for fuel cell application. Physica B: Condensed Matter. 2002;**323**:124-126. DOI: 10.1016/S0921-4526(02)00871-2

[16] Hui CL, Li XG, Hsing I-M. Well-dispersed surfactant-stabilized Pt/C nanocatalysts for fuel cell application: Dispersion control and surfactant removal. Electrochimica Acta. 2005;**51**:711-719. DOI: 10.1016/j.electacta.2005.05.024

[17] Li X, Hsing I-M. The effect of the Pt deposition method and the support on Pt dispersion on carbon nanotubes. Electrochimica Acta. 2006;**51**:5250-5258. DOI: 10.1016/j.electacta.2006.01.046

[18] Kou R, Shao Y, Mei D, Nie Z, Wang D, Wang CM, Viswanathan VV, Park SK, Akasay IA, Lin Y, Wang Y, Liu J. Stabilization of electrocatalytic metal nanoparticles at metal–metal oxide–graphene triple junction points. Journal of the American Chemical Society. 2011;**133**:2541-2547. DOI: 10.1021/ja107719u

[19] Xue T, Sun ZP, Wei L, Wang X, Lee J-M. One-step dual template synthesis of platinum on mesoporous carbon nanowires for electrocatalysts. International Journal of Hydrogen Energy. 2013;**38**:2754-2759. DOI: 10.1016/j.ijhydene.2012.12.034

[20] Neto AO, Giz MJ, Perez J, Ticianelli EA, Gonzalez ER. The electro-oxidation of ethanol on Pt-Ru and Pt-Mo particles supported on high-surface-area carbon. Journal of the Electrochemical Society. 2002;**149**:A272-A279. DOI: 10.1149/1.1446080

[21] Tayal J, Rawat B, Basu S. Bi-metallic and tri-metallic Pt-Sn/C, Pt-Ir-Sn/C catalysts for electro-oxidation of ethanol in direct ethanol fuel cell. International Journal of Hydrogen Energy. 2011;**36**:14884-14897. DOI: 10.1016/j.ijhydene.2011.03.035

[22] da Silva SG, Silva JCM, Buzzo GS, de Souza RFB, Spinacé EV, Neto AO, Assumpção MHMT. Electrochemical and fuel cell evaluation of PtAu/C electrocatalysts for ethanol electro-oxidation in alkaline media. International Journal of Hydrogen Energy. 2014;**39**:10121-10127. DOI: 10.1016/j.ijhydene.2014.04.169

[23] Pham VV, Ta V-T, Sunglae C. Synthesis of NiPt alloy nanoparticles by galvanic replacement method for direct ethanol fuel cell. International Journal of Hydrogen Energy. 2017;**42**:13192-13197. DOI: 10.1016/j.ijhydene.2017.01.236

[24] Sharma S, Pollet BG. Support materials for PEMFC and DMFC electrocatalysts–A review. Journal of Power Sources. 2012;**208**:96-119. DOI: 10.1016/j.jpowsour.2012.02.011

[25] Brownson DAC, Kampouris DK, Banks CE. An overview of graphene in energy production and storage applications. Journal of Power Sources. 2011;**196**:4873-4885. DOI: 10.1016/j.jpowsour.2011.02.022

[26] Zhang B, Fan L, Zhong H, Liu Y, Chen S. Graphene nanoelectrodes: Fabrication and size-dependent electrochemistry. Journal of the American Chemical Society. 2013;**135**:10073-10080. DOI: 10.1021/ja402456b

[27] Ambrosi A, Chua CK, Latiff NM, Loo AH, Wong CHA, Eng AYS, Binanni A, Pumera M. Graphene and its electrochemistry–An update. Chemical Society Reviews. 2016;**45**:2458-2493. DOI: 10.1039/C6CS00136J

[28] Ambrosi A, Chua CK, Bonanni A, Pumera M. Electrochemistry of graphene and related materials. Chemical Reviews. 2014;**114**:7150-7188. DOI: 10.1021/cr500023c

[29] Luo J, Kim J, Huang J. Material processing of chemically modified graphene: Some challenges and solutions. Accounts of Chemical Research. 2013;**46**:2225-2234. DOI: 10.1021/ar300180n

[30] Marinkas A, Arena F, Mitzel J, Prinz GM, Heinzel A, Peinecke V, Natter H. Graphene as catalyst support: The influences of carbon additives and catalyst preparation methods on the performance of PEM fuel cells. Carbon. 2013;**58**:139-150. DOI: 10.1016/j.carbon.2013.02.043

[31] Mu S, Chen X, Sun R, Liu X, Wu H, He D, Cheng K. Nano-size boron carbide intercalated graphene as high performance catalyst supports and electrodes for PEM fuel cells. Carbon. 2016;**103**:449-456. DOI: 10.1016/j.carbon.2016.03.044

[32] Sanli LI, Bayram V, Yarar B, Ghobadi S, Gürsel SA. Development of graphene supported platinum nanoparticles for polymer electrolyte membrane fuel cells: Effect of support type and impregnation–reduction methods. International Journal of Hydrogen Energy. 2016;**41**:3414-3427. DOI: 10.1016/j.ijhydene.2016.08.210

[33] Sanli LI, Bayram V, Yarar B, Ghobadi S, Güersel SA. Engineered catalyst layer design with graphene-carbon black hybrid supports for enhanced platinum utilization in PEM fuel cell. International Journal of Hydrogen Energy. 2017;**42**:1085-1092. DOI: 10.1016/j.ijhydene.2016.08.210

[34] Pinithchaisakula A, Themsirimongkon S, Promsawan N, Weankeaw P, Ounnunkad K, Saipanya S. An investigation of polydopamine-graphene oxide composite as a support for an anode fuel cell catalyst. Electrocatalysis. 2017;**8**:36-45. DOI: 10.1007/s12678-016-0338-6

[35] Cheng K, He D, Peng T, Lv H, Pan M, Mu S. Porous graphene supported Pt catalysts for proton exchange membrane fuel cells. Electrochimica Acta. 2014;**132**:356-363. DOI: 10.1016/j.electacta.2014.03.181

[36] Zhao D, Shi M-Q, Liu W-M, Chu Y-Q, Ma C-A. Special microwave-assisted one-pot synthesis of low loading Pt-Ru alloy nanoparticles on reduced graphene oxide for methanol oxidation. IET Micro & Nano Letters. 2014;**9**:50-54. DOI: 10.1049/mnl.2013.0525

[37] Nardecchia S, Carriazo D, Ferrer ML, Gutiérrez MC, del Monte F. Three dimensional macroporous architectures and aerogels built of carbon nanotubes and/or graphene: Synthesis and applications. Chemical Society Reviews. 2013;**42**:794-830. DOI: 10.1039/C2CS35353A

[38] Wei M, Qiao L, Zhang H, Karakalos S, Ma K, Fu Z, Swihart MT, Wu G. Engineering reduced graphene oxides with enhanced electrochemical properties through multiple-step reductions. Electrochimica Acta. 2017;**258**:735-743. DOI: 10.1016/j.electacta.2017.11.120

[39] Soo LT, Loh KS, Mohamad AB, Daud WRW, Wong WY. An overview of the electrochemical performance of modified graphene used as an electrocatalyst and as a catalyst support in fuel cells. Applied Catalysis A: Genereal. 2015;**497**:198-210. DOI: 10.1016/j.apcata.2015.03.008

[40] Tao L, Dou S, Ma Z, Shen AS, Wang S. Simultaneous Pt deposition and nitrogen doping of graphene as efficient and durable electrocatalysts for methanol oxidation. International Journal of Hydrogen Energy. 2015;**40**:14371-14377. DOI: 10.1016/j.ijhydene.2015.02.104

[41] Jafri RI, Rajalakshmi N, Dhathathreyan KS, Ramaprabhu S. Nitrogen doped graphene prepared by hydrothermal and thermal solid state methods as catalyst supports for fuel cell. International Journal of Hydrogen Energy. 2015;**40**:4347-4348. DOI: 10.1016/j.ijhydene.2015.02.008

[42] Jiao Y, Zheng Y, Davey K, Qiao S. Activity origin and catalyst design principles for electrocatalytic hydrogen evolution on heteroatom-doped graphene. Nature Energy. 2016;**1**:1-9. DOI: 10.1038/nenergy.2016.130

[43] Li Y, Li W, Ke T, Zhang P, Ren X, Deng L. Microwave assisted-synthesis of sulfur-doped graphene supported PdW nanoparticles as a high performance electrocatalyst for the oxygen reduction reaction. Electrochemistry Communications. 2016;**69**:68-71. DOI: 10.1016/j.elecom.2016.06.006

[44] Kwok YH, Tsang ACH, Wang Y, Leung DYC. Ultra-fine Pt nanoparticles on graphene aerogel as a porous electrode with high stability for microfluidic methanol fuel cell. Journal of Power Sources. 2017;**349**:75-83. DOI: 10.1016//j.jpowsour.2017.03.030

[45] Radhakrishnan T, Snadhyarani N. Three dimensional assembly of electrocatalytic platinum nanostructures on reduced graphene oxide–an electrochemical approach for high performance catalysts for methanol oxidation. International Journal of Hydrogen Energy. 2017;**42**:7014-7022. DOI: 10.1016/j.ijhydene.2016.12.132

[46] Qiu X, Li T, Deng S, Cen K, Xu L, Tang Y. A general strategy for the synthesis of PtM (M=Fe, co, Ni) decorated three-dimensional hollow graphene nanospheres for efficient methanol electrooxidation. Chemistry. 2018;**24**:1246-1252. DOI: 10.1002/chem.201704959

[47] Wang Y, Li Z, Xu S, Xie Y, Lin S. Highly dispersive platinum nanoparticles supporting on polyaniline modified three dimensional graphene and catalyzing methanol oxidation. Materials Research Bulletin. 2018;**102**:172-179. DOI: 10.1016/j.materresbull.2018.02.028

[48] Waenkaew P, Themsirimongkon S, Ounnunkad K, Promsawan N, Pinithchaisakula A, Saipanya S. Successive electrodeposition of polydopamine and PtPd metal on a graphene oxide support for use as anode fuel cell catalysts. Composite Interfaces. 2018;**25**:317-333. DOI: 10.1080/09276440.2018.1436798

[49] Hu C, Liu D, Xiao Y, Dai L. Functionalization of graphene materials by heteroatom-doping for energy conversion and storage. Progress in Natural Science: Materials International. DOI: 10.1016/j.pnsc.2018.02.001. (Forthcoming)

[50] Fan M, Feng Z-Q, Zhu C, Chen X, Chen C, Yang J, Sun D. Recent progress in 2D or 3D N-doped graphene synthesis and the characterizations, properties, and modulations of N species. 2016;**51**:10323-10349. DOI: 10.1007/s10853-016-0250-8

[51] Kanninen P, Luong ND, Sinh LH, Flórez-Montaño J, Jiang H, Pastor E, Seppälä J, Kallio T. Highly active platinum nanoparticles supported by nitrogen/sulfur functionalized graphene composite for ethanol electro-oxidation. Electrochimica Acta. 2017;**242**:315-326. DOI: 10.1016/j.electacta.2017.05.019

[52] Rethinasabapathy M, Kang S-M, Haldorai Y, Jankiraman M, Jonna N, Choe SR, Huh YS, Natesan B. Ternary PtRuFe nanoparticles supported on N-doped graphene as an efficient bifunctional catalyst for methanol oxidation and oxygen reduction reactions. International Journal of Hydrogen Energy. 2017;**42**:30738-30749. DOI: 10.1016/j.ijhydene.2017.10.121

[53] Zhao L, Sui X-L, Li J-Z, Zhang J-J, Zhang L-M, Huang G-S, Wang Z-B. Supramolecular assembly promoted synthesis of three-dimensional nitrogen doped graphene frameworks as efficient electrocatalyst for oxygen reduction reaction and methanol electrooxidation. Applied Catalysis B: Environmental. 2018;**231**:224-233. DOI: 10.1016/j.apcatb.2018.03.020

[54] Rybin M, Pereyaslavtsev A, Vasilieva T, Myasnikov V, Sokolov I, Pavlova A, Obraztsova E, Khomich A, Ralchenko V, Obraztsova E. Efficient nitrogen doping of graphene by plasma treatment. Carbon. 2016;**96**:196-202. DOI: 10.1016/j.carbon.2015.09.056

[55] He X, Yin F, Wang H, Chen B, Li G. Metal-organic frameworks for highly efficient oxygen electrocatalysis. Chinese Journal of Catalysis. 2018;**39**:207-227. DOI: 10.1016/S1872-2067(18)63017-7

[56] Jiang D, Cooper VR, Dai S. Porous graphene as the ultimate membrane for gas separation. Nano Letters. 2009;**9**:4019-4024. DOI: 10.1021/nI9021946

Chapter 8

Immobilization Impact of Photocatalysts onto Graphene Oxide

Ali Gemeay and Mohamed El-Halwagy

Additional information is available at the end of the chapter

http://dx.doi.org/10.5772/intechopen.78054

Abstract

The densely functionalized graphene oxide (GO) surface and the two-dimensional carbon structure had provided a unique opportunity for supporting photocatalysts. Concerning GO-based photocatalysis, GO plays the role of an electron acceptor that accelerates the interfacial electron-transfer process, recombination retardant of charge carriers, fine-tuner for the electronic and chemical properties of the supported photocatalysts, and finally, a carrier transport between different active sites. Moreover, standalone GO is a p-doped semiconductor material with the π^* orbital of the oxygen remains as the conduction band minimum (CBM) while the valance band maximum (VBM) changes gradually from the p-orbital of carbon to the 2p orbital of oxygen upon oxidation. The outstanding features of the GO-based photocatalysis opened the way to serve the progress in many environmental applications including water treatment, air purification, water splitting, CO_2 conversion, and sensing applications.

Keywords: photocatalysts, graphene oxide (GO), immobilization, environmental applications, energy conversion

1. Introduction

Over the past two decades, advanced oxidation processes (AOP) have received great attention, especially photocatalysis. The term photocatalysis was generally used to describe materials activated with light photons of appropriate energy to modify the rate of a chemical reaction without itself being transformed giving it the character of a catalyst. The photocatalytic process has provided an environmentally friendly method in many environmental and even non-environmental fields such as energy storage and conversion. However, the applications of photo-activated catalysts have been limited by the relatively low-efficiency resulting in a growing research activity to improve this technique.

The abundance of functional epoxide and hydroxyl groups on graphene oxide (GO) surfaces with the edges and rim sites around vacancies being decorated with pendant carboxylic acid, quinoidal, ketone, and lactone groups enables binding of active sites. Fortunately, the oxygenated groups can largely expand the structural/chemical diversity of GO by further chemical modification or functionalization, which offer an effective way to tailor the physical and chemical properties of GO to expected extents. Besides, GO also displays excellent optical and mechanical properties for a wide landscape of applications. Furthermore, the residual defects and holes arise through the reduction process of GO degrade the electronic quality of r-GO. As a consequence, GO, and GO-based composites have shown great potentials in the applications of energy storage/conversion and environment protection.

The properties of GO such as readily dispersible in water at the molecular level, biocompatibility, and tunable band gap motivated researchers to explore its potential as photocatalytic material. Furthermore, GO combines two complementary qualities, electrons imbibition, and consumption. The electrons imbibition property stimulates the interfacial electron-transfer process from TiO_2, actively limiting the chance of charge carriers recombination with a striking promoting of the photocatalytic response. Simultaneously, the consumption of the received electrons occurs during the partial reduction of GO to reconstruct the conjugated network of the graphene under ultraviolet (UV) assistance. This sensation leads to efficient charge separation and the possibility of more interactions between the composite and targeted organic compounds. In addition, TiO_2/GO composites extend the absorbable light range from the UV into the visible region.

2. Supported photocatalysts onto graphene oxide (GO)

Supported photocatalysts are a properly evolved concept in imparting progressed exposure of the catalysts to reactants and is common in industrial catalytic technologies. In this type of configuration, however, the nature of the photocatalyst-support interactions is important. For durable overall performance, a strong chemical bond is necessary, however, the influences of bonding on photocatalytic mechanisms are considered. The application of TiO_2 in suspension is effective in capturing sunlight due to the fact suspended TiO_2 powders have a high particular surface area within the range from 50 to 300 $m^2\ g^{-1}$, which in turn helps in keeping off mass transfer limitation, ensuing in a high photocatalytic activity. Moreover, TiO_2 alone showed a very low photocatalytic activity because of the rapid recombination of conduction band (CB) electrons and valence band (VB) holes. However, a light transport limitation appears with excessive catalyst loading. Besides, it is difficult to separate the small TiO_2 particles from the water after the remedy. To overcome this, the catalyst particles can be immobilized on a surface. In addition, this may lower the oxidation capacity in keeping with volume of water as compared to the suspension of solid particles system, due to the mass transfer difficulty and moderate transport limitation because of (i) a diminished catalyst surface-to-volume ratio, (ii) the presence of substrate that absorbs light and deteriorate its distribution, and (iii) a loss of movement of particles.

There are different kinds of materials that have been used as a support to fix photocatalysts. Among the different supports, GO is an excellent substrate material for many reasons due to its high specific surface area and superior electron mobility. There have been many efforts to immobilize TiO_2 photocatalyst over diverse structures of supports together with increasing the surface/volume ratio concurrently, which subsequently improves the photocatalytic oxidation efficiency. However, the surface area can only be productive if it allows efficient absorption of light. There is extensive attention to the preparation methods of GO-based material nanocomposites. GO-based material nanocomposites can be synthesized by numerous methods and approaches including the hydrothermal method, electrochemical co-deposition, in situ polymerization, microwave-assisted method, vacuum impregnation, and sol-gel technique. In the GO-based nanocomposite, GO presents either as a functional component or as a substrate for immobilizing the other components.

3. Heterogeneous GO-based photocatalytic materials

Research series conducted by Fujishima and Honda pull the trigger of scientific research of photocatalysis stimulation. Their initial demonstration was based on the activation of a semiconductor particulate material by the action of radiation with an appropriate wavelength to catalyze the dissociation of water. Since this time, several photocatalysts have been subject to extensive studies. In general, when a semiconductor photocatalyst material undergoes irradiation with a light of suitable wavelength, an electron gain a quantum of energy sufficient to its promotion to the CB. This electron transmission causes a positive hole in the VB. The electron in the CB and the hole in the VB are responsible for reduction or oxidation of any substrate, respectively, as shown in **Figure 1**. The role of graphene-metal oxide composites as photocatalysts, adsorbents, and disinfectants in water treatment was previously reviewed [1].

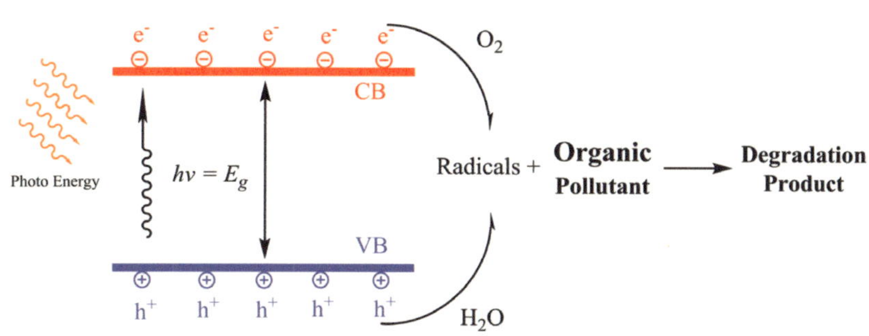

Figure 1. Simplification of photocatalysis mechanism.

3.1. Binary metal oxides

Titanium oxide (TiO_2) is the first and most important binary transition metal oxides that have been studied in the field of photocatalysis. TiO_2 is characterized by chemical stability and non-solubility in the aqueous medium, which facilitates the process of separation after the desired reactions. Moreover, photocatalytic stimulation is primarily aimed at environmental applications; therefore, non-toxicity of titanium dioxide is also favorable. Other n-type semiconductor binary oxides with a d-transition metal such as WO_3, Fe_2O_3, Nb_2O_5, V_2O_5, NiO, Ta_2O_5, ZrO_2, CuO, and Cu_2O have been studied. In addition, binary metal oxides of elements other than transition metals such as ZnO, Ga_2O_3, Sb_2O_3, Bi_2O_3, and CeO_2 had also some attention. Although ZnO suffers from photocorrosion, it is photocatalytic activity is comparable with TiO_2. Furthermore, the photocorrosion process can be controlled by monitoring of the operational factors such as the pH, additives, and ZnO crystal growth.

GO has been explored as electron acceptor molecule for making composite with TiO_2. The effects of particle size GO content and targeted pollutants for different TiO_2 and GO composites have been represented by examples in **Table 1**. Moreover, other metal oxides such as ZnO,

Composites	TiO_2 particle size	GO content	Pollutant	Ref.
Pt-GO-TiO_2/GR	30 nm	0.5 wt%	Dodecylbenzenesulfonate	[9]
TiO_2-GO and TiO_2-GR	5–15 nm	90 wt%	Methylene blue	[10]
TiO_2/GO	—	10 mg	Methylene blue	[11]
TiO_2GO	57 nm	3.3 wt%	Diphenhydramine methyl	[12]
TiO_2/GO	50 nm thickness	10 wt%	Methylene blue	[13]
TiO_2/GO	20–40nm	4.6 wt%	Methyl orange	[14]
TiO_2/GO	30 nm	10 wt%	Methylene blue	[15]
TiO_2/GO	—	GO:TiO_2 = 1.5 wt	Methyl orange	[16]
TiO_2/GO	10 nm	0.03 mg GO	Methylene blue	[17]
TiO_2/GO	4–5 nm	3.3–4.0 wt%	Diphenhydramine and methyl orange	[18]
TiO_2/GO	15 nm	~10%	Rhodamine B	[19]
TiO_2/GO	10 nm	GO:TiO_2 = 3:2 wt	Methyl orange	[20]
TiO_2/GO	6–9 nm	1 wt%	Methylene blue	[21]
Co_3O_4/TiO_2/GO	30–50 nm	3–4 wt%	Oxytetracycline and Congo Red	[22]
Fe_3O_4/TiO_2/GO	17 nm		Enrofloxacin	[23]
B/GO/TiO_2	51 nm		4-Nitrophenol	[24]
La/TiO_2/GO			Acid Blue 40	[25]

Table 1. Summary of TiO_2 and GO composites used as photocatalyst.

[2–4] NiO, [5] WO_3, [6] CuO, [7], and Mn_3O_4, [8] have been also incorporated with GO to serve as an enhanced photocatalyst.

3.2. Non-oxidic binary compounds

Many photocatalytic substances contain no oxygen, but still have another element of group 16 in the periodic table. Chalcogenide such as CdS, ZnS, Sb_2S_3, Bi_2S_3, MoS, CdSe, and CdTe give a good example of such materials. Sulfides other than zinc sulfide have an absorption edge compatible with solar energy utilization rather than photocatalytic processes. Unfortunately, sulfides have the disadvantage of photo-induced corrosion. Cadmium selenides and tellurides have a valence band redox potential lower than 1.23 V. This explains their poor photocatalytic activity for oxidation of organic pollutants [26].

A simple and high-yield room-temperature solid-state method was employed for the first time to fabricate GO/ZnS, GO/CdS, and GO/Bi_2S_3 composites [27]. The synthesized GO/metal sulfide composites were used as photocatalysts for the degradation of methyl orange under UV irradiation. The composites exhibited superior photocatalytic activity to pure metal sulfides, owing to the high specific surface area, and the reduction of photo-induced electron-hole pair recombination in metal sulfides due to the introduction of GO.

However, incorporation of non-oxidic binary compounds alone with GO is superfluous. They are often introduced to the GO surface with each other or with different photocatalytic materials. The association of ZnS-CdS with GO to form the ZnS-CdS/GO heterostructure for the photocatalytic H_2 gas generation result in the duplication of the recorded production rate. Moreover, doping ZnS-CdS/GO heterostructure with 2 wt% Pt nanoparticles to serve as co-catalysts, the hydrogen generation rate is significantly elevated to 1.68 and 0.78 mmol h^{-1} upon irradiation with UV-visible or visible light, respectively [28].

3.3. Multiple compounds

Metallates of elements of the middle region of the periodic table and cations with filled or partially filled orbitals in their outer shell, which may hybridize with the oxygen 2p orbitals. Aluminates, ferrites, niobates, tantalates, titanates, tungstates, and vanadates are some examples of these metallates. While the examples of those cations are 6 s orbitals in Pb^{2+} and Bi^{3+}, and 5 s in Sn^{2+} or 4d in Ag^{+}. Various ternary and quaternary oxides, oxynitrides, oxysulfides, and oxyhalides have also been investigated for evaluation of their activity in the degradation of organic pollutants and water splitting.

Incorporation of metallates with GO has also been investigated as an enhanced photocatalyst. For instance, the Bi_2WO_6/GO photocatalysts were successfully prepared via in situ refluxing method in the presence of GO. The photocatalytic degradation of rhodamine B under visible light irradiation was utilized to evaluate the photocatalytic performance. The photocatalytic activity of Bi_2WO_6/graphene was greatly enhanced compared to pure Bi_2WO_6. The enhanced photocatalytic activity could be attributed to that GO as charge transfer channel [29].

4. Photocatalytic roles of graphene oxide (GO)

4.1. Intrinsic photocatalytic activity

Most of the photocatalytic materials are wide band gap semiconductors. GO is a p-doped material because oxygen atoms are more electronegative than carbon atoms [30]. The band gap of GO can be tunable by just varying the oxidation level. The fully oxidized GO can act as an electrical insulator while the partially oxidized GO can act as a semiconductor [31]. Introducing more oxygen enlarges the band gap, and the valence band maximum (VBM) gradually changes from the p-orbital of graphene to the 2p orbital of oxygen; the π^* orbital remains as the conduction band minimum (CBM), **Figure 2**.

The photocatalytic characteristics of GO nanostructures were confirmed by measuring reduction rate of resazurin into resorufin as a function of UV irradiation time [32]. The band gap of GO locates between 2.4 and 4.3 eV. Upon excitation, the electron-hole pairs will be generated at the GO surface. Hence, the localized defects on the GO nanosheets trap the produced electrons and limit their recombination process with the counter-currently produced holes.

4.2. Photocatalyst support

Since GO was rediscovered in 2004, it represents the top of carbon materials in many of its properties including outstanding electronic, thermal, and mechanical properties. GO as a support has many distinctive features including two-dimensional structure with the large specific surface area, high adsorption capacity, and excellent dispersibility in both aqueous and organic solvents [33]. Furthermore, the spread of the oxygenated functional groups facilitates the decoration of its surface with photoactive spots. As GO can add many advantages to the photocatalyst, it can also avoid many special problems related to the active site. The photocorrosion of ZnO was inhibited to a large extent upon incorporation with GO [34].

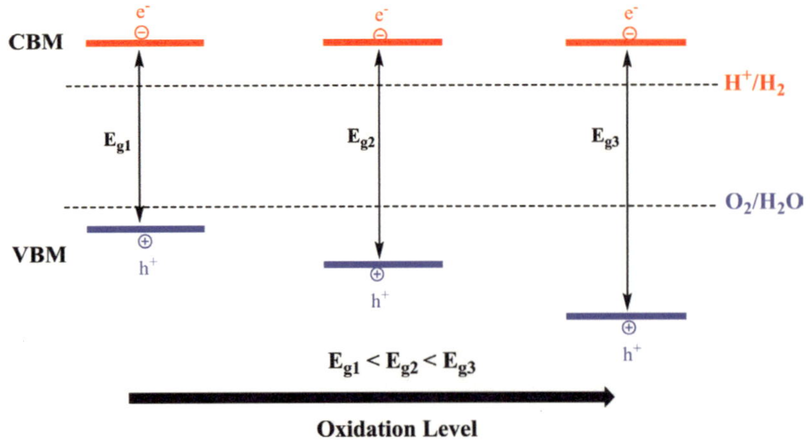

Figure 2. The VBM and CBM positions of GO sheets with varying oxidation level.

http://dx.doi.org/10.5772/intechopen.78054

We cannot overlook some of the constraints that may limit the use of GO in supporting photocatalysts. The first thing to issue is the operating temperature range. The temperature between 150 and 200°C is usually sufficient for the start of the GO labile oxygen-containing groups decomposition [35]. The second issue that there is a claim that the $OH^{\bullet}$ radicals generated at the TiO_2 surface could lead to an oxidative attack on the carbon-rich structure, which is ultimately mineralized as evident from the decreased total organic carbon (TOC) concentration with increasing irradiation time [36].

4.3. Charge separation and gap control

It is expected that the chemical bonding and associated charge transfer at the interface between the photocatalyst and GO support can be used to fine-tune the electronic and chemical properties of the active sites. For instance, TiO_2 is the widely used photocatalyst, however, its activity is restricted to the ultraviolet region because of its wide band gap, and the photo-generated electron-hole pairs in experience rapid recombination [37, 38]. GO sheets are particularly effective in separating charges on TiO_2 as the photo-generated electrons will move toward the GO support, leaving behind the holes in the supported semiconductor particles. Thus, the GO support acts as an electron acceptor to enhance the separation between the photo-generated electron and holes, thereby suppressing their recombination, and consequently enhancing the photocatalytic efficiency. In addition, the unpaired π-electrons on GO can interact with TiO_2 to extend the light absorption range of TiO_2 (**Figure 3**). The GO/TiO_2 composites exhibited excellent photochemical activity under visible light irradiation.

The photocatalytic activity of the gap controlled composite is governed by other factors including GO to active sites ratio. After the threshold limit, the photocatalytic activity decrease with increasing the absorption and scattering of photons by the excess carbon content in the composite. Photocatalytic activity of hybrid material also depends on the interface between TiO_2 and graphene an intense coupling between TiO_2 and GO facilitates charge separation and so retards recombination. Liang et al. [19] reported the growth of uniform TiO_2 nanocrystals directly on GO substrate via hydrolysis with subsequent hydrothermal treatment.

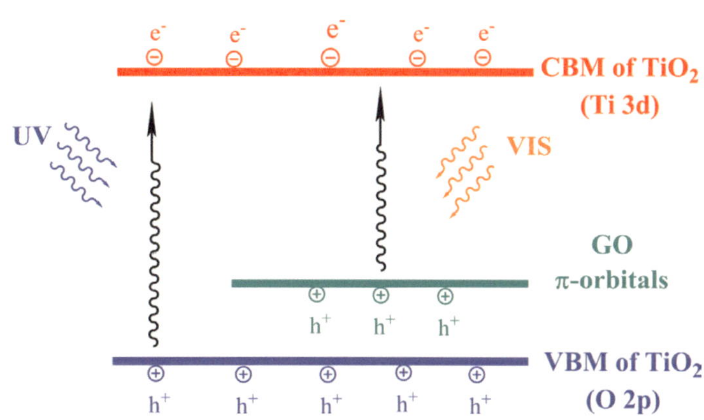

Figure 3. The interaction between unpaired π-electrons on GO with surface TiO_2.

The resultant GO/TiO_2 hybrids were tested in the photocatalytic degradation of the well-known rhodamine B dye. In comparison with commercial P25 TiO_2, the prepared hybrids showed a three time faster degradation rate due to the enhanced electronic combination between GO and TiO_2 in addition to the remarkable higher gross surface area.

4.4. Heterojunction formation

Incorporating GO with metal-containing semiconductors can initiate a p-n junction, which considerably improves the separation of photo-generated charges. This is a possible way to fabricate GO/semiconductor composites with different properties by using a tunable semiconductor [39]. GO/TiO_2 composites were prepared by using $TiCl_3$ and GO as reactants. Again, the concentration of GO in starting solution played an important role in the photocatalytic performance of the composites. The heterojunction between p-type GO and n-type TiO_2 semiconductors functioned as the separator for the photo-generated electron-hole pairs, **Figure 4**. These semiconductors could be excited by visible light with wavelengths longer than 510 nm for the degradation of methyl orange. Also, the integration of GO with TiO_2 will significantly increase the photovoltaic response and significantly prolong its mean life-time of electron-hole pairs compared with that of TiO_2 [40]. Similar to TiO_2, ZnO also can form a p-n heterojunction with GO for visible light absorption. Quantum dot sized ZnO nanoparticles deposited on graphene sheets with a p-n heterojunction interface were demonstrated by a change of photocurrent direction at different bias potential that significantly enhanced photoresponse properties under solar light irradiation [41].

4.5. Coupling multiple active sites

GO can act as a common platform for more than one active site to produce enhanced heterostructure for photocatalytic activity. For example, ZnS-CdS/GO shows a twice activity toward photocatalytic hydrogen generation compared with ZnS-CdS standalone heterostructures [28].

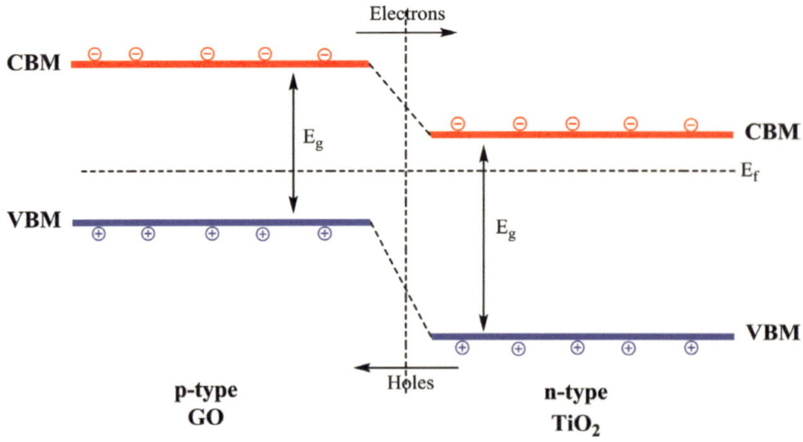

Figure 4. p–n heterojunction formation at the interface between GO and TiO2.

Hence, GO construct a carrier transport channel between ZnS and CdS to enhance cooperative effects. In addition, GO oxide platform allow the loading of 2 wt% Pt nanoparticles as co-catalysts for the ZnS-CdS/GO heterostructures. Under UV-visible and visible light irradiation the hydrogen generation rate of the resultant heterostructure is significantly improved to 1.68 and 0.78 mmol h^{-1}, respectively.

5. Environmental applications

5.1. Water treatment

The contamination problem of natural water resources with pollutants of different forms is a problem that threatens public health. The photocatalysis is expected to play a major role in the water purification process because it has a great ability to exploit solar energy in pollutants degradation in the moderate temperature range. Visible light-responsive photocatalysis has been widely investigated for the treatment of inorganic, organic, and biological contaminated water. However, the application of photocatalysis in water treatment is still in the research stage.

5.1.1. Degradation of organic pollutants

Since organic dyes pharmaceuticals are usually leakage with a significant part to the industrial wastewater during manufacture processes; it has received a lot of attention in terms of treatment processes, including photocatalysis. Because of the easy tracing of dye decolonization process, it has left a great scientific legacy of published scientific papers. In general, when the photocatalyst is irradiated with photon with energy compatible with the band gap energy (Eg), an electron is transferred from the VB to the CB, leaving behind a hole. Accordingly, the produced electron-hole pairs are involved in a series of chain oxidative-reductive reactions, Eqs. (1)–(10) [42].

$$\text{Photocatalyst} + h\nu \longrightarrow \text{h}^+ + \text{e}^- \quad (1)$$

$$\text{h}^+ + \text{H}_2\text{O} \longrightarrow {}^\bullet\text{OH} + \text{H}^+ \quad (2)$$

$$\text{h}^+ + \text{OH}^- \longrightarrow {}^\bullet\text{OH} \quad (3)$$

$$\text{h}^+ + \text{pollutant} \longrightarrow (\text{pollutant})^+ \quad (4)$$

$$\text{e}^- + \text{O}_2 \longrightarrow {}^\bullet\text{O}_2^- \quad (5)$$

$$\cdot\text{O}_2^- + \text{H}^+ \longrightarrow {}^\bullet\text{OOH} \quad (6)$$

$$2\,^{\bullet}OOH \longrightarrow O_2 + H_2O_2 \quad (7)$$

$$H_2O_2 + {^{\bullet}O_2^-} \longrightarrow {^{\bullet}OH} + OH^- + O_2 \quad (8)$$

$$H_2O_2 + h\nu \longrightarrow 2\,^{\bullet}OH \quad (9)$$

$$\text{pollutant} + ({^{\bullet}OH}, h^+, {^{\bullet}OOH} \text{ or } {^{\bullet}O_2^-}) \longrightarrow \text{degradation product} \quad (10)$$

Recently, enhanced photocatalytic ZnO/GO nanocomposite was introduced for the degradation of methyl orange under UV irradiation [43]. The reaction proceeds for 2 h to achieve 95% degradation from the initial dye concentration of 10 mg/L. Pseudo-first-order kinetics were recorded with respect to methyl orange with an apparent reaction rate constants in the range of 0.009–0.030 min^{-1}. Moreover, the aforementioned ZnO/GO photocatalyst passed the stability and reusability test.

5.1.2. Stalemate of inorganic pollutants

Chromium is one of the common cationic pollutants in wastewater. Chromium exists in two main oxidation states in the environment, Cr(VI) and Cr(III). Compared with Cr(III), Cr(VI) is more toxic [44]. Hexagonal nanoflowers of Tin(IV) sulfide/GO was synthesized by hydrothermal method [45]. The composite showed a high photocatalytic activity toward the oxidation of hexavalent chromium upon contact with the SnS_2/GO under visible light irradiation, Eqs. (11)–(13).

$$SnS_2/GO + h\nu \longrightarrow h^+ + e^- \quad (11)$$

$$2\,h^+ + H_2O \longrightarrow \tfrac{1}{2}O_2 + 2\,H^+ \quad (12)$$

$$CrO_4^{2-} + 8\,H^+ + 3\,e^- \longrightarrow Cr^{3+} + 4\,H_2O \quad (13)$$

5.1.3. Membrane purification

Membrane fouling and poor water flux are the main problems facing water treatment by membrane technique. However, incorporation of photocatalysis onto the membrane surface is expected to increase the water flux as result of the photo-enhanced hydrophilicity and contaminant degradation. Surface modification by photocatalyst grafting provides a very promising route to the fabrication of high-performance photocatalytic membranes for sustainable water treatment.

Deposition of TiO_2 and GO nanosheets on the polysulfone-base membrane was one of the first attempts to construct a photocatalytic membrane [46]. Several properties of the TiO_2/GO/

polysulfone membrane over pristine polysulfone membrane have been identified, including the effective photodegradation of methylene blue under UV (about 60–80% faster) and sunlight (3–4 times faster) beside improved membrane flux. After that many attempts have been carried out to construct and utilized photocatalytic membrane for water and salty water purification [47–51]. In the next-generation of water treatment photocatalytic membranes, the antimicrobial character was also included. In situ implementation of Ag nanoparticles onto the surface of GO/TiO_2 nanocomposite membrane assemblies, allowing for simultaneous filtration, and disinfection [52].

5.2. Air pollutants removal

According to national ambient air quality standards (NAAQS) carbon monoxide (CO), ozone (O_3), particulate matter, nitrogen oxides (NO_x, x = 1 or 2), sulfur dioxide (SO_2), and lead (Pb) considered harmful to public health and the environment. Unfortunately, gases such as NOx, SO_2, and CO are difficult to oxidize and remove. Hence, the role of photocatalysis as an effective economic alternative to oxidize the aforementioned gases to products, which are harmless to public health.

Recently, novel TiO_2/GO functionalized with a cobalt complex for significant degradation of NO_x and CO has been reported [53]. The cobalt-imidazole functionalized GO incorporating with the TiO_2 result in a notable decrease of the band gap and increased the sensitivity to visible light irradiation. Experimental results revealed a high percentage deterioration of NOx (51%) and CO (46%). This photocatalyst reflects the ability of TiO_2 to act as photocatalyst with the ability of GO to modify its band gap with the exploitation of cobalt to form complex formation with pollutant gases.

5.3. CO_2 reduction

Reduction of carbon dioxide (CO_2) to hydrocarbons represents a perfect example of simultaneous solar energy harvesting. Side-by-side to exploit CO_2, the main cause of global warming, emitted into the atmosphere as a major product of combustion operations. Unfortunately, exploiting CO_2 gas needs high energy. Therefore, the existence of some processes related to the exploitation of solar energy as an energy source represents an opportunity for two birds with one stone, that is, the energy harvesting/storage and environmental issues. Superior advantages of the graphene-based photocatalyst for CO_2 reduction was categorized into six aspects [54]: (i) prevent carrier recombination, (ii) increasing specific surface areas, (iii) strong π-π interaction between graphene and CO_2, (iv) enhancing photocatalyst mechanical and chemical stability, (v) improving nanoparticles (NPs) dispersion, and (vi) enhancing light absorption.

Since GO could act as an active photocatalyst, it was used as a photocatalyst for CO_2 to methanol conversion. The photo-reduction of CO_2 to methanol conversion rate on modified GO achieve six-fold higher than the pure TiO_2 under 300-watt halogen lamp irradiation [55]. The research was not limited to use GO standalone, but there is a growing effort to improve using TiO_2 spread over the surface of the GO. For instance, GO doped oxygen-rich TiO_2 hybrid

(GO-OTiO$_2$) was synthesized through the wet chemical impregnation technique [56]. The photocatalytic performances were evaluated through the photoreduction of CO_2 under the irradiation of low-power energy-saving daylight bulbs. GO extent the reactivity of doped oxygen-rich TiO_2 to 95.8% even after 6 h of light irradiation. This observation firmly established the role of GO as an effective catalyst support. The composite with 5 wt%, GO has a CH_4 yield of 1.718 μmol g^{-1}cat. after 6 h of reaction, that is, 14.0 folds comparison to commercial Degussa P25.

Besides TiO_2, other semiconductors, clusters, nanoparticles, and complexes immobilized onto GO has also been investigated for CO_2 reduction to hydrocarbon. Hexamolybdenum cluster compound was used for the modification of GO for visible light absorption [57]. After 24 h of visible light illumination, the yield of methanol was found to be 1644 and 1294 μmol g^{-1}cat. for GO/$Cs_2Mo_6Br^i_8Br^a_x$ and GO/$(TBA)_2Mo_6Br^i_8Br^a_x$ (TBA = tetrabutylammonium), respectively. These values are much higher than GO alone (439 μmol g^{-1}cat.). Halide perovskite quantum dots ($CsPbBr_3$ QD/GO) was utilized for the photochemical conversion of CO_2 [58]. These materials primarily regarded as optoelectronic materials for light emitting diode (LED) and photovoltaic devices. Compared to the individual $CsPbBr_3$ QDs, the rate of electron consumption improved from 23.7 to 29.8 μmol g^{-1}cat. after the incorporation with the GO. As an example of nanoparticles decorated GO with a series of Cu-nanoparticles were utilized to reduce CO_2 to hydrocarbon fuels under visible light irradiation [59]. Cu-NPs (10 wt%) effectively tune the work function of GO 60 times higher. A ruthenium trinuclear polyazine complex was also immobilized onto GO functionalized with phenanthroline ligands [60]. Using a 20 W white cold LED flood light, in a dimethyl formamide-water mixture sacrificing agent, the yield of methanol was found to be 3977.57 ± 5.60 μmol g^{-1}cat. after 48 h.

5.4. Water splitting

Water splitting with hydrogen evolution based on photocatalytic process represents a renewable energy production with no reliance on fossil fuels and no CO_2 emission. The role of GO in photocatalytic water splitting was perfectly reviewed [61]. GO itself as a photocatalyst steadily catalyzes H_2 generation from 20 vol% aqueous methanol solution and pure water under irradiation with UV or visible light [62]. Concurrently with the photocatalytic reaction, the CB of GO with an excessive overpotential associated with the level for H_2 generation reveals fast electron injection from the excited GO into the solution. On the other hand, the holes in the VB of GO do not interact with the water molecules for O_2 generation but are disabled to the hole scavenger (methanol) alternatively.

The GO band gap decrease during the photocatalytic reaction because of GO reduction. However, the H_2 evolution remained unchanged, indicating that the involved oxygenated sites remain accessible to protons in the liquid phase.

Recently, GO sheets covalently functionalized with (5, 10, 15, 20-tetraphenyl) porphinato manganese(III) (MnTPP) GO-MnTPP and tested for water splitting under UV-vis light irradiation [63]. The dye moiety acted as a sensitizer while GO moiety acted as a supporting matrix, electron mediator, enhancing photoexcited electrons transfer, and suppressing charges recombination. Under UV-vis irradiation, the GO-MnTPP hybrid demonstrates considerable photocatalytic activity for H_2 evolution.

5.5. Sensing applications

Photoelectrochemical sensors based on materials with photoelectric properties, such as TiO_2 have drawn attention in bioanalysis, medical diagnoses, and pollutant monitoring. A density functional theory (DFT) study was carried out to model the NO_2 interaction with pristine and N-doped TiO_2/GO nanocomposites for a promising NO_2 gas sensor [64]. Another example the construction of a blue-light photoelectrochemical sensor based on nickel tetra-amine phthalocyanine-GO for the detection of erythromycin [65]. The prepared composite was fixed on the surface of the indium tin oxide coated glass electrode. When the light was incident on the surface of the material the electron-hole pairs generation process started with a photocurrent response. The erythromycin suppresses electron/hole recombination resulted in changes in the photocurrent with a linear response for erythromycin concentrations ranging from 0.40 to 120.00 $\mu mol\ L^{-1}$.

6. Future perspectives

Incorporation of GO with different semiconductor, metal nanostructures, and complexes allows the design of many generations of photocatalyst systems. In addition, great opportunities still exist in the exploitation of novel GO base hybrids and composites photocatalyst. Dual function photocatalysis that can perform multitasks simultaneously represents the greatest future trend along with the discovery of new materials. This is not a fantasy; previously a dual-functional photocatalysis was introduced based on a ternary hybrid of TiO_2 modified with GO along with Pt and fluoride for both H_2-producing water treatment and degradation of organic pollutants [66]. The positive effect of GO on the dual photocatalytic activity was observed only when Pt and surface fluoride were co-present.

7. Conclusion

Photocatalysts alone almost showed a very low photocatalytic activity because of the rapid recombination of CB electrons and VB holes. The chemical bonding and associated charge transfer at the interface between the photocatalyst and GO support can be used to fine-tune the electronic and chemical properties of the active sites. GO can act as a common platform for more than one active site to produce enhanced heterostructure for photocatalytic activity.

GO is an excellent supporting material due to its high specific surface area and superior electron mobility. GO plays the role of an electron acceptor that accelerates the interfacial electron-transfer process from photocatalysts materials, which strongly hindering the recombination of charge carriers and thus improving the photocatalytic activity. The spread of the oxygenated functional groups on its surface facilitates the process of planting photoactive spots on its surface.

The band gap of GO is tunable by just varying the oxidation level. Fully oxidized GO act as an electrical insulator and partially oxidized GO can act as a semiconductor. Introducing more oxygen enlarges the band gap, and the VBM gradually changes from the p-orbital of graphene to the 2p orbital of oxygen; the π^* orbital remains as the CBM.

Aluminates, ferrites, niobates, tantalates, titanates, tungstates, and vanadates are examples of the metallates, which are incorporated with GO and revealed an enhanced photocatalytic activity on the degradation of organic dyes under visible light due to charge transfer channel of GO.

The heterojunction between p-type GO and n-type semiconductors functioned as the separator for the photo-generated electron-hole pairs. These semiconductors could be excited by visible light with wavelengths longer than 510 nm for the degradation of methyl orange. In most cases, GO served as an electron sink to facilitate separation and store the separated electrons.

Surface modification by photocatalyst grafting provides a very promising route to the fabrication of high-performance photocatalytic membranes for sustainable water treatment.

Some of the outstanding properties of GO-based photocatalysts in CO_2 reduction processes have been shortlisted: (i) block carrier recombination, (ii) Improving specific surface areas, (iii) strong π–π interaction with CO_2, (iv) enhancing either photocatalyst mechanical and chemical stability, (v) improving nanoparticles dispersion, (vi) intensifying light absorption.

Author details

Ali Gemeay[1]* and Mohamed El-Halwagy[2]

*Address all correspondence to: agemeay@science.tanta.edu.eg

1 Chemistry Department, Faculty of Science, Tanta University, Tanta, Egypt

2 Sidi Kerir Petrochemicals Co., Alexandria, Egypt

References

[1] Upadhyay RK, Soin N, Roy SS. Role of graphene/metal oxide composites as photocatalysts, adsorbents and disinfectants in water treatment: A review. RSC Advances. 2014;**4**:3823. DOI: 10.1039/c3ra45013a

[2] Li B, Liu T, Wang Y, Wang Z. ZnO/graphene-oxide nanocomposite with remarkably enhanced visible-light-driven photocatalytic performance. Journal of Colloid and Interface Science. 2012;**377**:114-121. DOI: 10.1016/j.jcis.2012.03.060

[3] Ameen S, Shaheer Akhtar M, Seo H-K, Shik Shin H. Advanced ZnO–graphene oxide nanohybrid and its photocatalytic applications. Materials Letters. 2013;**100**:261-265. DOI: 10.1016/j.matlet.2013.03.012

[4] Liu S, Sun H, Suvorova A, Wang S. One-pot hydrothermal synthesis of ZnO-reduced graphene oxide composites using Zn powders for enhanced photocatalysis. Chemical Engineering Journal. 2013;**229**:533-539. DOI: 10.1016/j.cej.2013.06.063

[5] Agegnehu AK, Pan C-J, Rick J, Lee J-F, Su W-N, Hwang B-J. Enhanced hydrogen generation by cocatalytic Ni and NiO nanoparticles loaded on graphene oxide sheets. Journal of Materials Chemistry. 2012;**22**:13849. DOI: 10.1039/c2jm30474k

[6] Zhao Y, Wei X, Wang Y, Luo F. One-pot twelve tungsten phosphate acid assisted electrochemical synthesis of WO_3-decorated graphene sheets for high-efficiency UV-light-driven photocatalysis. Chemical Physics Letters. 2014;**607**:34-38. DOI: 10.1016/j.cplett.2014.05.043

[7] Huang T, Wu J, Zhao Z, Zeng T, Zhang J, Xu A, et al. Synthesis and photocatalytic performance of CuO-CeO_2/graphene oxide. Materials Letters. 2016;**185**:503-506. DOI: 10.1016/j.matlet.2016.09.057

[8] Catalysis A. In situ fabrication of Mn_3O_4 decorated graphene oxide as a synergistic catalyst for degradation of methylene blue. Applied Catalysis B : Environmental. 2015;**162**:268-274. DOI: 10.1016/j.apcatb.2014.06.058

[9] Neppolian B, Bruno A, Bianchi CL, Ashokkumar M. Graphene oxide based Pt–TiO_2 photocatalyst: Ultrasound assisted synthesis, characterization and catalytic efficiency. Ultrasonics Sonochemistry. 2012;**19**:9-15. DOI: 10.1016/j.ultsonch.2011.05.018

[10] Ismail AA, Geioushy RA, Bouzid H, Al-Sayari SA, Al-Hajry A, Bahnemann DW. TiO_2 decoration of graphene layers for highly efficient photocatalyst: Impact of calcination at different gas atmosphere on photocatalytic efficiency. Applied Catalysis B: Environmental. 2013;**129**:62-70. DOI: 10.1016/j.apcatb.2012.09.024

[11] Min Y, Zhang K, Zhao W, Zheng F, Chen Y, Zhang Y. Enhanced chemical interaction between TiO_2 and graphene oxide for photocatalytic decolorization of methylene blue. Chemical Engineering Journal. 2012;**193-194**:203-210. DOI: 10.1016/j.cej.2012.04.047

[12] Pastrana-Martíne LM, Morales-Torres S, Kontos AG, Moustakas NG, Faria JL, Doña-Rodríguez JM, et al. TiO_2, surface modified TiO_2 and graphene oxide-TiO_2 photocatalysts for degradation of water pollutants under near-UV/Vis and visible light. Chemical Engineering Journal. 2013;**224**:17-23. DOI: 10.1016/j.cej.2012.11.040

[13] Zhang YP, Xu JJ, Sun ZH, Li CZ, Pan CX. Preparation of graphene and TiO_2 layer by layer composite with highly photocatalytic efficiency. Progress in Natural Science: Materials International. 2011;**21**:467-471. DOI: 10.1016/S1002-0071(12)60084-7

[14] Jiang G, Lin Z, Chen C, Zhu L, Chang Q, Wang N, et al. TiO_2 nanoparticles assembled on graphene oxide nanosheets with high photocatalytic activity for removal of pollutants. Carbon. 2011;**49**:2693-2701. DOI: 10.1016/j.carbon.2011.02.059

[15] Nguyen-Phan TD, Pham VH, Shin EW, Pham HD, Kim S, Chung JS, et al. The role of graphene oxide content on the adsorption-enhanced photocatalysis of titanium dioxide/graphene oxide composites. Chemical Engineering Journal. 2011;**170**:226-232. DOI: 10.1016/j.cej.2011.03.060

[16] Pu X, Zhang D, Gao Y, Shao X, Ding G, Li S, et al. One-pot microwave-assisted combustion synthesis of graphene oxide-TiO_2 hybrids for photodegradation of methyl orange. Journal of Alloys and Compounds. 2013;**551**:382-388. DOI: 10.1016/j.jallcom.2012.11.028

[17] Yoo D-H, Cuong TV, Pham VH, Chung JS, Khoa NT, Kim EJ, et al. Enhanced photocatalytic activity of graphene oxide decorated on TiO_2 films under UV and visible irradiation. Current Applied Physics. 2011;**11**:805-808. DOI: 10.1016/j.cap.2010.11.077

[18] Pastrana-Martínez LM, Morales-Torres S, Likodimos V, Figueiredo JL, Faria JL, Falaras P, et al. Advanced nanostructured photocatalysts based on reduced graphene oxide–TiO_2 composites for degradation of diphenhydramine pharmaceutical and methyl orange dye. Applied Catalysis B: Environmental. 2012;**123-124**:241-256. DOI: 10.1016/j.apcatb.2012.04.045

[19] Liang Y, Wang H, Sanchez Casalongue H, Chen Z, Dai H. TiO_2 nanocrystals grown on graphene as advanced photocatalytic hybrid materials. Nano Research. 2010;**3**:701-705. DOI: 10.1007/s12274-010-0033-5

[20] Gao Y, Pu X, Zhang D, Ding G, Shao X, Ma J. Combustion synthesis of graphene oxide–TiO_2 hybrid materials for photodegradation of methyl orange. Carbon. 2012;**50**:4093-4101. DOI: 10.1016/j.carbon.2012.04.057

[21] Kim SP, Choi HC. Photocatalytic degradation of methylene blue in presence of graphene oxide/TiO_2 nanocomposites. Bulletin of the Korean Chemical Society. 2014;**35**:2660-2664. DOI: 10.5012/bkcs.2014.35.9.2660

[22] Jo W-K, Kumar S, Isaacs MA, Lee AF, Karthikeyan S. Cobalt promoted TiO_2/GO for the photocatalytic degradation of oxytetracycline and Congo Red. Applied Catalysis B: Environmental. 2017;**201**:159-168. DOI: 10.1016/j.apcatb.2016.08.022

[23] Yu Y, Yan L, Cheng J, Jing C. Mechanistic insights into TiO_2 thickness in Fe_3O_4@TiO_2-GO composites for enrofloxacin photodegradation. Chemical Engineering Journal. 2017;**325**:647-654. DOI: 10.1016/j.cej.2017.05.092

[24] Shokri A, Mahanpoor K, Soodbar D. Evaluation of a modified TiO_2(GO-B-TiO_2) photo catalyst for degradation of 4-nitrophenol in petrochemical wastewater by response surface methodology based on the central composite design. Journal of Environmental Chemical Engineering. 2016;**4**:585-598. DOI: 10.1016/j.jece.2015.11.007

[25] Oppong SOB, Anku W, Shukla KS, Govender P. Lanthanum doped-TiO_2 decorated on graphene oxide nanocomposite: A photocatalyst for enhanced degradation of acid blue 40 under simulated solar light. Advanced Materials Letters 2017;**8**:295-302. DOI: 10.5185/amlett.2017.6826

[26] Hernández-Ramírez A, Medina-Ramírez I. Photocatalytic Semiconductors: Synthesis, Characterization, and Environmental Applications. 1st ed. Cham: Springer; 2015. 289 p. DOI: 10.1007/978-3-319-10999-2

[27] Chen F-J, Cao Y-L, Jia D-Z. A room-temperature solid-state route for the synthesis of graphene oxide–metal sulfide composites with excellent photocatalytic activity. CrystEngComm. 2013;**15**:4747. DOI: 10.1039/c3ce40079d

[28] Wang X, Yuan B, Xie Z, Wang D, Zhang R. ZnS–CdS/Graphene oxide heterostructures prepared by a light irradiation-assisted method for effective photocatalytic hydrogen generation. Journal of Colloid and Interface Science. 2015;**446**:150-154. DOI: 10.1016/j.jcis.2015.01.051

[29] Min Y-L, Zhang K, Chen Y-C, Zhang Y-G. Enhanced photocatalytic performance of Bi_2WO_6 by graphene supporter as charge transfer channel. Separation and Purification Technology. 2012;**86**:98-105. DOI: 10.1016/j.seppur.2011.10.025

[30] Wang X, Li X, Zhang L, Yoon Y, Weber PK, Wang H, et al. N-doping of graphene through electrothermal reactions with ammonia. Science. 2009;**324**:768-771. DOI: 10.1126/science.1170335

[31] Loh KP, Bao Q, Eda G, Chhowalla M. Graphene oxide as a chemically tunable platform for optical applications. Nature Chemistry. 2010;**2**:1015-1024. DOI: 10.1038/nchem.907

[32] Krishnamoorthy K, Mohan R, Kim S-J. Graphene oxide as a photocatalytic material. Applied Physics Letters. 2011;**98**:244101. DOI: 10.1063/1.3599453

[33] Paredes JI, Marti A, Tasco JMD, Martı A. Graphene oxide dispersions in organic solvents graphene oxide dispersions in organic solvents. 2008;**24**:10560-4. DOI: 10.1021/la801744a

[34] Fan H, Zhao X, Yang J, Shan X, Yang L, Zhang Y, et al. ZnO-graphene composite for photocatalytic degradation of methylene blue dye. Catalysis Communications. 2012;**29**:29-34. DOI: 10.1016/j.catcom.2012.09.013

[35] Gemeay AH, El-Halwagy ME, El-Sharkawy RG, Zaki AB. Chelation mode impact of copper(II)-aminosilane complexes immobilized onto graphene oxide as an oxidative catalyst. Journal of Environmental Chemical Engineering. 2017;**5**:2761-2772. DOI: 10.1016/j.jece.2017.05.020

[36] Radich JG, Krenselewski AL, Zhu J, Kamat PV. Is graphene a stable platform for photocatalysis? Mineralization of reduced graphene oxide with UV-irradiated TiO_2 nanoparticles. Chemistry of Materials. 2014;**26**:4662-4668. DOI: 10.1021/cm5026552

[37] Robert D. Photosensitization of TiO_2 by M_xO_y and M_xS_y nanoparticles for heterogeneous photocatalysis applications. Catalysis Today. 2007;**122**:20-26. DOI: 10.1016/j.cattod.2007.01.060

[38] Yin W-J, Chen S, Yang J-H, Gong X-G, Yan Y, Wei S-H. Effective band gap narrowing of anatase TiO_2 by strain along a soft crystal direction. Applied Physics Letters. 2010;**96**:221901. DOI: 10.1063/1.3430005

[39] Chen C, Cai W, Long M, Zhou B, Wu Y, Wu D, et al. Synthesis of visible-light responsive graphene oxide/TiO_2 composites with p/n heterojunction. ACS Nano. 2010;**4**:6425-6432. DOI: 10.1021/nn102130m

[40] Wang P, Zhai Y, Wang D, Dong S. Synthesis of reduced graphene oxide-anatase TiO_2 nanocomposite and its improved photo-induced charge transfer properties. Nanoscale. 2011;**3**:1640. DOI: 10.1039/c0nr00714e

[41] Min Y, Zhang K, Chen L, Chen Y, Zhang Y. Ionic liquid assisting synthesis of ZnO/graphene heterostructure photocatalysts with tunable photoresponse properties. Diamond and Related Materials. 2012;**26**:32-38. DOI: 10.1016/j.diamond.2012.04.003

[42] Dong S, Feng J, Fan M, Pi Y, Hu L, Han X, et al. Recent developments in heterogeneous photocatalytic water treatment using visible light-responsive photocatalysts: A review. RSC Advances. 2015;**5**:14610-14630. DOI: 10.1039/C4RA13734E

[43] Nguyen VN, Tran DT, Nguyen MT, Le TTT, Ha MN, Nguyen MV, et al. Enhanced photocatalytic degradation of methyl orange using ZnO/graphene oxide nanocomposites. Research on Chemical Intermediates. 2018;**3081**:3095-3144. DOI: 10.1007/s11164-018-3294-3

[44] Liu SS, Chen YZ, De Zhang L, Hua GM, Xu W, Li N, et al. Enhanced removal of trace Cr(VI) ions from aqueous solution by titanium oxide–Ag composite adsorbents. Journal of Hazardous Materials. 2011;**190**:723-728. DOI: 10.1016/j.jhazmat.2011.03.114

[45] Han L, Mao D, Huang Y, Zheng L, Yuan Y, Su Y, et al. Fabrication of unique tin(IV) sulfide/graphene oxide for photocatalytically treating chromium(VI)-containing wastewater. Journal of Cleaner Production. 2017;**168**:519-525. DOI: 10.1016/j.jclepro.2017.09.027

[46] Gao Y, Hu M, Mi B. Membrane surface modification with TiO_2–graphene oxide for enhanced photocatalytic performance. Journal of Membrane Science. 2014;**455**:349-356. DOI: 10.1016/ j.memsci.2014.01.011

[47] Xu C, Xu Y, Zhu J. Photocatalytic antifouling graphene oxide-mediated hierarchical filtration membranes with potential applications on water purification. ACS Applied Materials & Interfaces. 2014;**6**:16117-16123. DOI: 10.1021/am5040945

[48] Pastrana-Martínez LM, Morales-Torres S, Figueiredo JL, Faria JL, Silva AMT. Graphene oxide based ultrafiltration membranes for photocatalytic degradation of organic pollutants in salty water. Water Research. 2015;**77**:179-190. DOI: 10.1016/j.watres.2015.03.014

[49] Xu Z, Wu T, Shi J, Teng K, Wang W, Ma M, et al. Photocatalytic antifouling PVDF ultrafiltration membranes based on synergy of graphene oxide and TiO_2 for water treatment. Journal of Membrane Science. 2016;**520**:281-293. DOI: 10.1016/j.memsci.2016.07.060

[50] Rao G, Zhang Q, Zhao H, Chen J, Li Y. Novel titanium dioxide/iron (III) oxide/graphene oxide photocatalytic membrane for enhanced humic acid removal from water. Chemical Engineering Journal. 2016;**302**:633-640. DOI: 10.1016/j.cej.2016.05.095

[51] Liu G, Han K, Ye H, Zhu C, Gao Y, Liu Y, et al. Graphene oxide/triethanolamine modified titanate nanowires as photocatalytic membrane for water treatment. Chemical Engineering Journal. 2017;**320**:74-80. DOI: 10.1016/j.cej.2017.03.024

[52] Jiang Y, Liu D, Cho M, Lee SS, Zhang F, Biswas P, et al. In situ photocatalytic synthesis of Ag nanoparticles (nAg) by crumpled graphene oxide composite membranes for filtration and disinfection applications. Environmental Science & Technology. 2016;**50**: 2514-2521. DOI: 10.1021/acs.est.5b04584

[53] Seifvand N, Kowsari E. Novel TiO_2/graphene oxide functionalized with a cobalt complex for significant degradation of NOx and CO. RSC Advances. 2015;**5**:93706-93716. DOI: 10.1039/C5RA13620B

[54] Low J, Yu J, Ho W. Graphene-based photocatalysts for CO_2 reduction to solar fuel. Journal of Physical Chemistry Letters. 2015;**6**:4244-4251. DOI: 10.1021/acs.jpclett.5b01610

[55] Hsu H-C, Shown I, Wei H-Y, Chang Y-C, Du H-Y, Lin Y-G, et al. Graphene oxide as a promising photocatalyst for CO_2 to methanol conversion. Nanoscale. 2013;**5**:262-268. DOI: 10.1039/C2NR31718D

[56] Tan L-L, Ong W-J, Chai S-P, Goh BT, Mohamed AR. Visible-light-active oxygen-rich TiO_2 decorated 2D graphene oxide with enhanced photocatalytic activity toward carbon dioxide reduction. Applied Catalysis B: Environmental. 2015;**179**:160-170. DOI: 10.1016/j.apcatb.2015.05.024

[57] Kumar P, Mungse HP, Cordier S, Boukherroub R, Khatri OP, Jain SL. Hexamolybdenum clusters supported on graphene oxide: Visible-light induced photocatalytic reduction of carbon dioxide into methanol. Carbon. 2015;**94**:91-100. DOI: 10.1016/j.carbon.2015.06.029

[58] Xu Y-F, Yang M-Z, Chen B-X, Wang X-D, Chen H-Y, Kuang D-B, et al. A CsPbBr 3 perovskite quantum dot/graphene oxide composite for photocatalytic CO_2 reduction. Journal of the American Chemical Society. 2017;**139**:5660-5663. DOI: 10.1021/jacs.7b00489

[59] Shown I, Hsu H-C, Chang Y-C, Lin C-H, Roy PK, Ganguly A, et al. Highly efficient visible light photocatalytic reduction of CO_2 to hydrocarbon fuels by cu-nanoparticle decorated graphene oxide. Nano Letters. 2014;**14**:6097-6103. DOI: 10.1021/nl503609v

[60] Kumar P, Sain B, Jain SL. Photocatalytic reduction of carbon dioxide to methanol using a ruthenium trinuclear polyazine complex immobilized on graphene oxide under visible light irradiation. Journal of Materials Chemistry A. 2014;**2**:11246. DOI: 10.1039/c4ta01494d

[61] Yeh T-F, Cihlář J, Chang C-Y, Cheng C, Teng H. Roles of graphene oxide in photocatalytic water splitting. Materials Today. 2013;**16**:78-84. DOI: 10.1016/j.mattod.2013.03.006

[62] Yeh T-F, Syu J-M, Cheng C, Chang T-H, Teng H. Graphite oxide as a photocatalyst for hydrogen production from water. Advanced Functional Materials. 2010;**20**:2255-2262. DOI: 10.1002/adfm.201000274

[63] Li X, Li K, Wang D, Huang J, Zhang C, Du Y, et al. One-pot synthesis of manganese porphyrin covalently functionalized graphene oxide for enhanced photocatalytic hydrogen evolution. Journal of Porphyrins and Phthalocyanines. 2017;**21**:179-188. DOI: 10.1142/S1088424616501236

[64] Abbasi A, Sardroodi JJ. TiO_2/graphene oxide heterostructures for gas-sensing: Interaction of nitrogen dioxide with the pristine and nitrogen modified nanostructures investigated by DFT. Surface Review and Letters. 2018;**1850170**:1850170. DOI: 10.1142/s0218625x18501706

[65] Peng J, Huang Q, Zhuge W, Liu Y, Zhang C, Yang W, et al. Blue-light photoelectrochemical sensor based on nickel tetra-amined phthalocyanine-graphene oxide covalent compound for ultrasensitive detection of erythromycin. Biosensors & Bioelectronics. 2018;**106**:212-218. DOI: 10.1016/j.bios.2018.02.009

[66] Cho Y-J, Kim H, Lee S, Choi W. Dual-functional photocatalysis using a ternary hybrid of TiO_2 modified with graphene oxide along with Pt and fluoride for H_2-producing water treatment. Journal of Catalysis. 2015;**330**:387-395. DOI: 10.1016/j.jcat.2015.07.007

Chapter 9

Chemically Exfoliated Graphene Nanosheets for Flexible Electrode Applications

Joong Tark Han, Seung Yol Jeong,
Hee Jin Jeong and Geon-Woong Lee

Additional information is available at the end of the chapter

http://dx.doi.org/10.5772/intechopen.77284

Abstract

Graphene oxide (GO), produced by oxidation of graphite powder and exfoliation, is intensively utilized in electrodes, templates for hybrid materials, interfacial modifiers, three-dimensional structures, and so on, with its performance as an electrode material being determined by its chemical and structural states. This chapter describes the fabrication method of GO nanosheets from graphite oxide powder and their stable dispersion after reduction and applications in devices. Rheologically driven exfoliation and unusual acoustic cavitation methods were applied to produce large and less defective GO nanosheets. As a dispersion strategy of reduced GO (RGO) in solution, TiO_2 precursor, cation-π interaction, silanol groups were introduced. Moreover, supramolecular chemistry, for example, quadruple hydrogen bonding moieties, was applied to solve the dispersion of highly concentrated RGO pastes. As potential applications of GO and RGO, we described GO as a p-type dopant and interfacial modifier as well as energy storage electrodes, IR sensors, and emitters. The judicious use of chemically exfoliated graphene can open new applications as a flexible electrode.

Keywords: graphene oxide, reduced graphene oxide, exfoliation, dispersion

1. Introduction

Atomically thin graphene oxide (GO), produced by oxidation and exfoliation of graphite powder, has been intensively studied for applications in electrodes, templates for hybrid materials, interfacial modifiers, three-dimensional structures, and so on [1–4]. Its performance as an electrode material is determined by its chemical and structural states. The topological defects present in the basal plane of reduced GO (RGO) can significantly affect its electrical and electrochemical

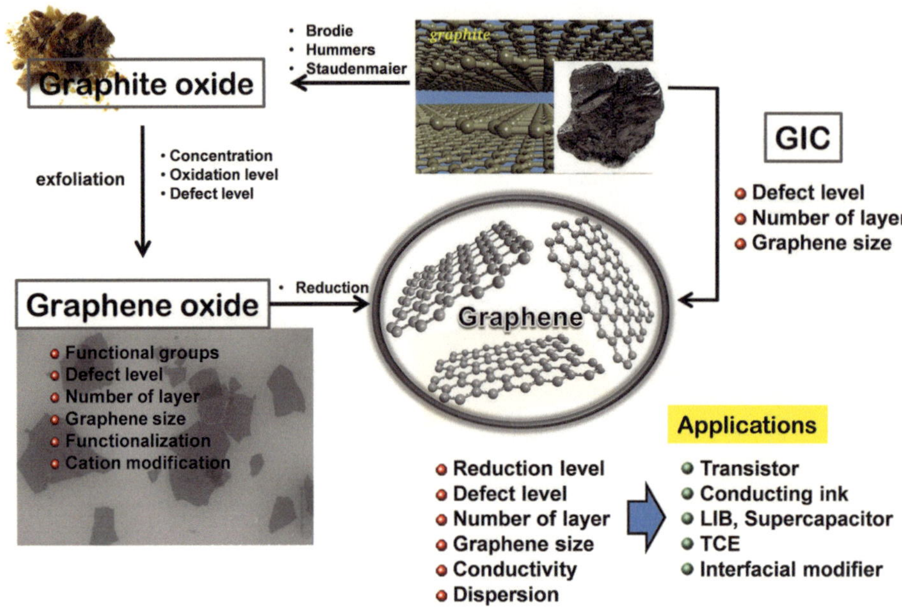

Figure 1. Production schematics of chemically exfoliated graphene from graphite. Here, GIC stands for graphite intercalation compound and LIB for lithium ion batteries.

properties. As shown in **Figure 1**, GO nanosheets are typically produced by oxidizing graphite using strong acids and oxidants, followed by exfoliation in aqueous solutions [5–8]. It should be noted that, the characteristics of GO and RGO nanosheets critically depend on the oxidation states of graphite oxide and its exfoliation. Moreover, for real-life applications, the dispersion stability of RGO inks or pastes is a prerequisite. The dispersion of high-quality chemically exfoliated graphene (CEG) or RGO in polar solvents, which contain few oxygen functional groups and defects, has been impossible due to the hydrophobic nature of graphene without post treatment or addition of dispersant molecules. The stability of RGO dispersion is one of the crucial factors for preserving their unique properties such as electrical conductivity and mechanical strength.

Therefore, this chapter describes some of the research on CEG nanosheets conducted over the past 8 years that addresses these and other challenges, with an emphasis on our own efforts. We began with the efficient fabrication method of single layer GO nanosheets from graphite, and then described the stable dispersion of RGO in solutions. Furthermore, we described the applications of GOs as p-type dopants, conductors and interfacial modifiers. We concluded with some discussion of future directions and the remaining challenges in chemically exfoliated graphene technologies.

2. Efficient fabrication of single layer graphene oxide nanosheets

2.1. Rheologically driven exfoliation of graphite oxide

Conventional sonicators vigorously destroy the structure of GO, which results in producing small-sized GO nanosheets due to acoustical wave agitation in solution. An alternative way to minimize the destructive effect of exfoliation of graphite oxide is to use a homogenizer to apply a shear force in the solution (**Figure 2(A)**). The average lateral size of sonicated GO (SGO) nanosheets (a few square micrometers) was smaller than that of the homogenized GO (HGO) nanosheets (a few hundred square micrometers) in the optical images in **Figure 2(B)** and **(C)**. The SGO nanosheets exhibited some agglomerated GO on the silicon substrate due to the small size distribution of the sheets. To confirm the exfoliation effects of HGO and SGO sheets, we carried out homogeneous dispersion of GO sheets in aqueous solution without using small size graphite powder (70 μm). The rheologically derived or sonicated exfoliation and dispersion of GO sheets was accomplished (**Figure 1(d)**) in an aqueous NaOH solution at pH 10 for 1 h. After diluting it using dimethylformamide (DMF), the RGO solution was prepared by chemical reduction with hydrazine for preparing transparent conducting films. The enhanced sheet resistance of the reduced HGO (HRGO) thin film was found to be 2.2 kΩ/sq. at 80% transmittance. The effective exfoliation method has great potential for application for high performance GO-based flexible electrodes.

2.2. Extremely efficient liquid exfoliation of graphite oxide using unusual acoustic cavitation

Acoustic cavitation, also called sonication, has been used to fabricate two-dimensional (2-D) nanosheets via exfoliation of bulk-layered crystal materials in solution to fabricate fascinating materials such as graphene, transition metal dichalcogenides, and transition metal oxides. The

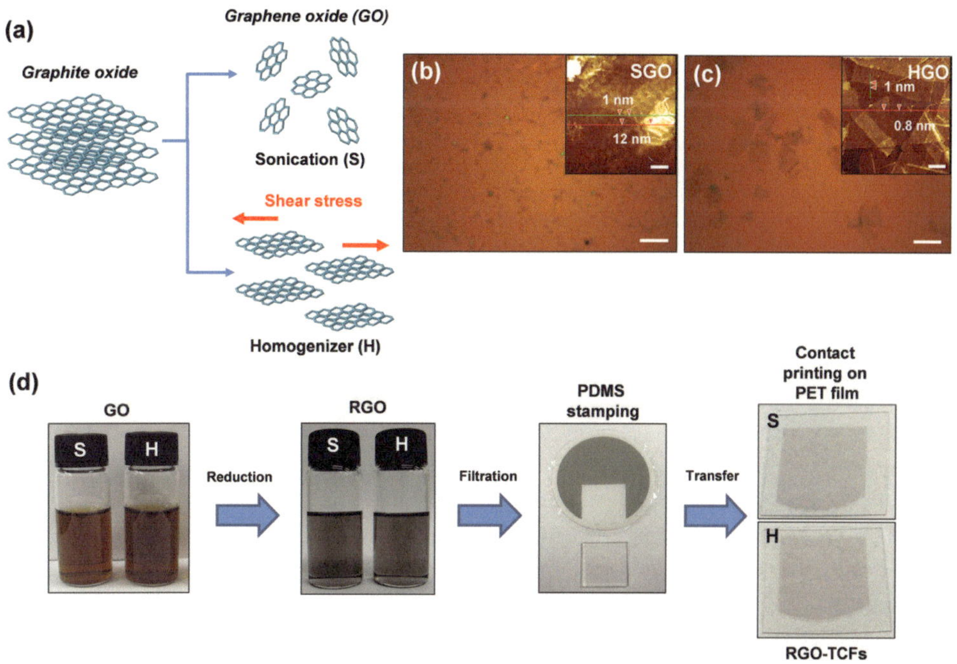

Figure 2. Fabrication of GO and RGO nanosheets by using different exfoliation methods. (A) Exfoliation of graphite oxide by sonication (S) or homogenization (H). (B) and (C) optical images of GO samples prepared by sonication (B) and homogenization (C) deposited on a 300-nm-thick SiO2 substrate (inset: AFM images of the GO sheets). (D) Fabrication of transparent conducting films (TCFs) with RGO nanosheets by the contact printing of filtrated RGO films. Here PDMS is polydimethylsiloxane, PET is polyethylene terephthalate, and RGO-TCFs are reduced graphene oxide transparent conductive films. The scale bars in (B) and (C) are 10 μm, and those in the respective insets are 2 μm [9].

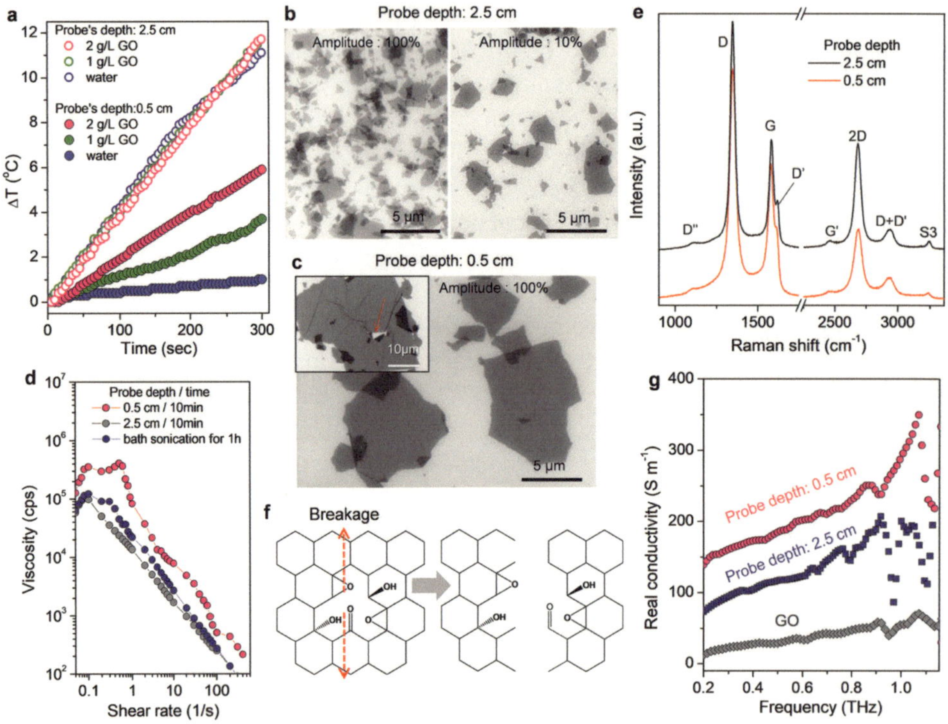

Figure 3. (a) Temperature change (ΔT) over time during sonication of pure water and of a GO suspension containing different initial amounts of graphite oxide. (b) and (c) FESEM images of GO nanosheets fabricated using probe sonication by dipping probe into the liquid surface by 2.5 and 0.5 cm, respectively, for 10 min. The large GO nanosheet was fabricated at 0.5 probe depth condition. (d) Shear viscosity of GO paste samples showing different rheological behavior due to their sizes. (e) Raman spectra of the chemically reduced GO nanosheets demonstrating the effect of acoustic cavitation at different probe depth on the crystalline structure of RGO (f) breakage of GO nanosheet initiated at the defect site or from the sp^3 hybridized region during probe sonication. (g) Real THz conductivity of the GO, SRGO, and LRGO films prepared by sonication with probe depths of 0.5 and 2.5 cm [13].

high energetic transient acoustic cavitation; the formation, growth, and implosive collapse of bubbles at high ultrasonic intensities (10–30 W cm^{-2}) in a liquid medium, allow to give physical effects on exfoliation of layered materials. However, the high energetic transient cavitation phenomenon can give a detrimental effect on 2D materials by generating defects on the surface, which decrease their electrical and other useful properties. Recently large (>10 μm) chemically modified graphene nanosheets have been developed from graphite oxide. These have fewer defects than those produced by other methods without requiring further separation processes and can be produced by combining ultrasonic acoustic cavitation with sufficient acoustic shearing and additional microbubbling by aeration in an extremely short time (10 min). It can be achieved by adjusting the ultrasound parameters (amplitude, time, and probe immersion depth) for the delivered power (related to temperature change (ΔT) **Figure 3(a)**), the acoustic flow rate, and the bubbling behavior in 200 mL water using conventional flat tip probes with a 12.7 mm diameter. In order to reduce the detrimental effect of transient cavitation, the probe tip was located at a 0.5 cm depth. Subsequently, the acoustic flow rate decreased from 0.62 to 0.47 m s^{-1} and then increased to 0.73 m s^{-1} at 100% amplitude, which was faster than the 10%

amplitude at the 2.5 cm probe depth. Moreover, bubbling due to the liquid surface instability under acoustic oscillation is also helpful for the efficient exfoliation of graphite oxide. Bubbling by aeration at the liquid surface is also helpful for the dispersion of nanomaterials because bubbling can produce a greater shearing effect on the particles in suspension under an acoustic flow with lower energy. **Figure 3(b)** and **(c)** shows the scanning electron microscopy (SEM) images of exfoliated GO nanosheets under different cavitation conditions. The maximum size of GO was dramatically increased by adjusting the probe depth from 2.5 to 0.5 cm. At a probe depth of 2.5 cm, the lateral size of GO was less than 5 μm even at 10% amplitude after 10 min of sonication (**Figure 3(b)**) because of breakage in the stretched C–C or C–O–C bonds [10, 11] due to the high energetic physical phenomena of microjets and shock waves [12]. However, at a probe depth of 0.5 cm, GO nanosheets with a maximum 30 μm size were produced even at the high output power setting (amplitude 100%) by reducing the detrimental effect of the high energy cavitation process (**Figure 3(c)**).

3. Dispersion of reduced graphene oxide

3.1. TiO_2 precursor-assisted dispersion of reduced graphene oxide in solution

The problems associated with the aggregation of the RGO sheets in organic solvents were addressed by introducing noncovalent interactions among the sp^2 carbons of the RGO sheets and the TiO_2 precursor sol, as shown in **Figure 4(a)**. Titanium dioxide is also a promising charge screening candidate because it can interact electrostatically with oxygen moieties causing charge trapping [14, 15]. The TiO_2 precursor sol was prepared from a titanium isopropoxide (TIP)/acac stabilizer (1/5 molar ratio) solution, which was added to the GO solution. In order to determine the minimum amount of used TiO_2 precursor for RGO dispersion, the varying amount of TiO_2 precursor sols were added into the GO solution prior to hydrazine reduction. The weight ratio between GO and TIP in the precursor TiO_2 sol was varied between 0 and 1.5. Just a 0.1 weight ratio was required to stabilize the RGO solution in dimethylformamide (DMF) after hydrazine reduction. This stable RGO/TiO_2 precursor sol mixture can be deposited onto the large area substrate by air-spraying without postreduction process. Usually, RGO films are fabricated by deposition of GO nanosheets on the substratem, followed by thermal or chemical reduction at elevated temperatures. Moreover, direct deposition of RGO solution onto the substrate induces formation of wrinkled structures, which can decrease their electrical properties. However, wrinkle-free RGO/TiO_2 hybrid multilayer films can be built up on SiO_2 by automatic spray-coating. The electrical transport characteristics of the RGO and RGO/TiO_2 hybrid films were investigated by preparing graphene field-effect transistors (FETs) on heavily doped Si substrates, which are commonly employed as gate electrodes. It is worth noting that the conductivity of the RGO film at the neutral charge point was maximized for GO/TIP (1/0.7 ratio) by inserting a thin TiO_2 layer into the RGO multilayer film, despite the amorphous insulating characteristics of TiO_2. This was due to the hole-doping effect caused by increasing the TiO_2 amount between the RGO nanosheets, which was demonstrated by observing a significant blue shift of the G peaks in Raman spectra.

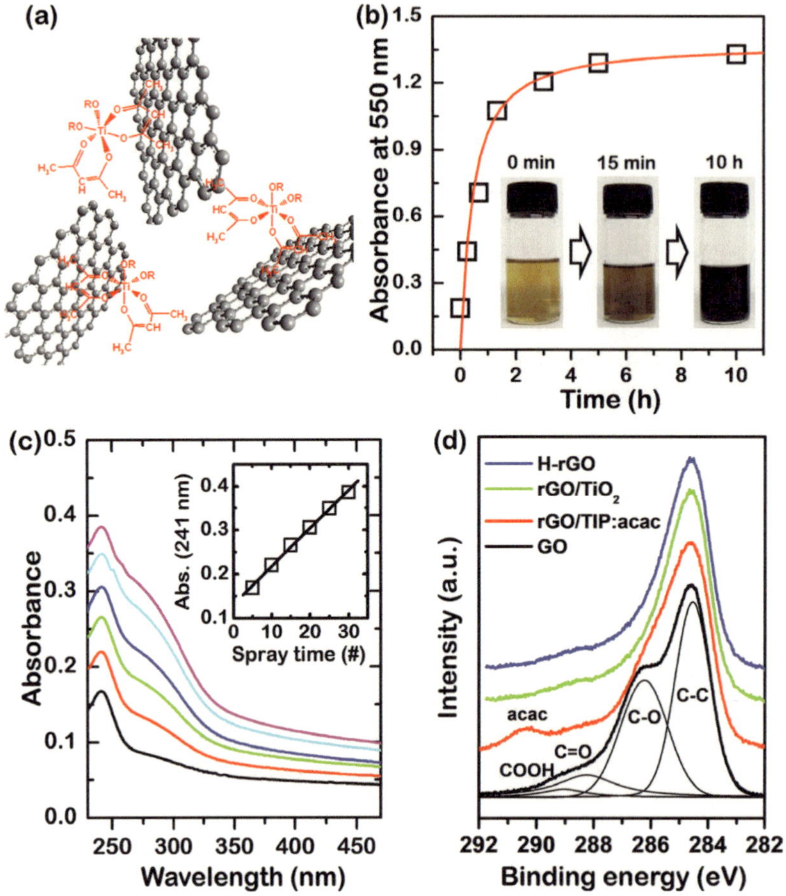

Figure 4. (a) Proposed mechanism for the dispersion of RGO sheets by the TiO_2 precursor sol via a hydrophobic interaction. (b) Dispersion stability of RGO solution in DMF after chemical reduction with hydrazine monohydrate; gradual increase of absorbance of GO solution at 550 nm and vial images shows the stable dispersion of RGO dispersion. (c) UV–Vis absorption spectra of RGO/TiO_2 hybrid multilayer films; the linear increase of absorption intensity shows the regular deposition of films by spraying. (d) C1s XPS spectra of GO, RGO reduced by hydrazine vapor (H-RGO), RGO/TIP:acac (TiO_2 precursor), and RGO/TiO_2 hybrid film thermally treated at 200°C [16].

3.2. Dispersion of reduced graphene oxide nanosheets by monovalent cation-π interaction

The cation-π interaction on crystallized RGO, which has fewer defects and oxygen functional groups, can be enhanced the dispersion stability in various solvents due to Coulombic repulsion between the cations on the in-plane of graphene. (Jeong et al. [17]) **Figure 5** shows the stable dispersion of RGO by monovalent cation–π interactions. The interactions did not directly occur on the basal plane of GO because the cations usually interact with the oxygen functional groups of highly oxidized GO (as described in **Figure 5(a)–(c)**). In order to effectively activate the cation interaction on π stage, the sp^2 carbon state on the basal plane is exposed through the reduction process (**Figure 5(a)**, **(b)**, **(d)**, and **(e)**). Therefore, mild reduction and aging processes are necessary to increase the six-membered sp^2 carbon states as described in **Figure 5(d)**.

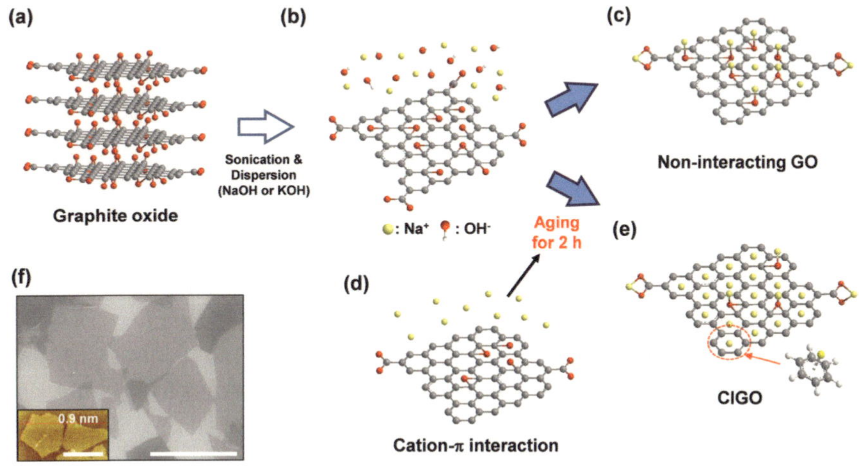

Figure 5. (a) Structure of graphite oxide, (b) as-exfoliated GO in NaOH solution, (c) highly interacting cations with oxygen functional groups. (a, b, d, e) procedure for obtaining the cation–π interacting GO. (d) Intermediate state of GO by mild deoxygenation aging in NaOH solution (e) decoration of cations on the partially reduced GO surface via a cation-π interaction. (f) FE-SEM image of CIGO on a silicon wafer (inset: AFM images of CIGO sheets). The scale bars in (f) and the inset are 2 μm [17].

Here, the electrostatic binding enthalpy of cations to a π system ($-\Delta H = 19.2$ kcal/mol) was higher than that that of water ($-\Delta H = 17.9$ kcal/mol). As the aging time is optimized, GO can be formed due to the cation interacting GO (CIGO) as shown in **Figure 5(e)**. Interestingly, the dispersion stability of noninteracting GO and CIGO were similar in aqueous solutions because of the presence of oxygen functional groups. However, the significant differences occurred after chemical reduction, which is described in **Figure 5(c)** and **(e)**. These results show that the cations with interacting sp^2 carbon did not desorb from the CIGO after the reduction process. Following the hydrazine reduction, the CIGO formed a dispersion of cation–π interacting RGO (CIRGO), whereas, the noninteracting GO aggregated in aqueous solution, as shown in **Figure 5(c)**. Moreover, the atomic force microscopy (AFM) image of the single-layered CIGO in **Figure 5(f)** confirmed its 2 μm size and 0.9 nm thickness.

3.3. Spontaneous reduction and dispersion of graphene oxide nanosheets with *in-situ* synthesized hydrazine assisted by hexamethyldisilazane

For real-life applications of RGO nanosheets, alcohol-based formulations of graphene are sometimes needed for graphene processing if the use of harsh organic solvents is not possible. Alcoholic solvents are not good for the dispersion of RGO in solutions due to their solubility parameters. Therefore, for the stable dispersion of RGO in alcoholic solvents, dispersant molecules should be added before the chemical reduction of GO in solutions. Recently, it has been reported that hexamethyldisilazane (HMDS) is a good candidate for the dispersion of RGO in alcohol because HMDS can be easily hydrolyzed into trimethylsilanol and ammonia in the presence of water. Furthermore, for the reduction of GO in solutions, hydrazine can be *in-situ* synthesized in a GO suspension by mimicking a typical reaction cycle involving GO (using alternative ketone molecules as catalysts) and ammonia and hydrogen peroxide as reagents.

Thus, HMDS can be used as a source of ammonia molecules for synthesizing hydrazine and dispersing RGO (**Figure 6**). The step-wise heating of the solution at 50 and 100°C is required to utilize keton groups in GO for in-situ synthesis of hydrazine molecules at high temperature for reduction of GO.

3.4. Dispersant-free dispersion of reduced graphene oxide by supramolecular chemistry

Highly concentrated colloidal suspensions of graphene nanosheets are of great interest for a variety of applications ranging from flexible electronics and conducting fibers to electrochemical electrodes for energy harvesting or storage devices. Unfortunately, many additives such as organic surfactants and polymeric dispersants should be added to prepare highly concentrated graphene pastes. These organic dispersant molecules can give detrimental effects on their electrical or thermal properties because graphene nanosheets can be separated by insulating organic materials if it is removed. Quadruple hydrogen bond (QHB) networks can overcome these issues for fabricating printable, spinnable, and chemically compatible conducting pastes containing high quality graphene nanosheets in organic solvents without the need for additional dispersion agents. Motivated by the self-assembly of donor-donor-acceptor-acceptor (DDAA) arrays of hydrogen bonding sites, GO nanosheets were functionalized

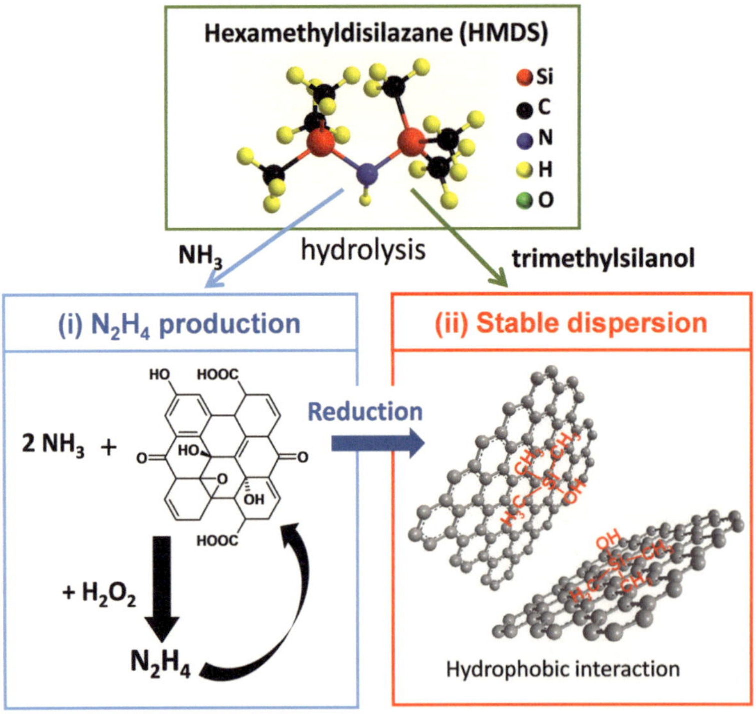

Figure 6. Roles of hexamethyldisilazane (HMDS): (i) ammonia source for the GO-assisted production of hydrazine upon the addition of hydrogen peroxide and (ii) RGO dispersion agent in ethanol, via hydrophobic interactions [18].

http://dx.doi.org/10.5772/intechopen.77284

using 2-ureido-4[1*H*]pyrimidinone (UHP) moieties to provide QHB motifs (**Figure 7(a)**). QHB arrays are much stronger than triple hydrogen bond arrays and are easily accessible synthetically. **Figure 7(b)** shows the well-dispersed RGO paste in DMF illustrating the striking synergy effect of QHB moieties into graphene nanosheets on the fabrication of dispersant-free RGO pastes. This unique paste can be used in electrochemical and printed electrodes and could be formed into flexible buckypaper.

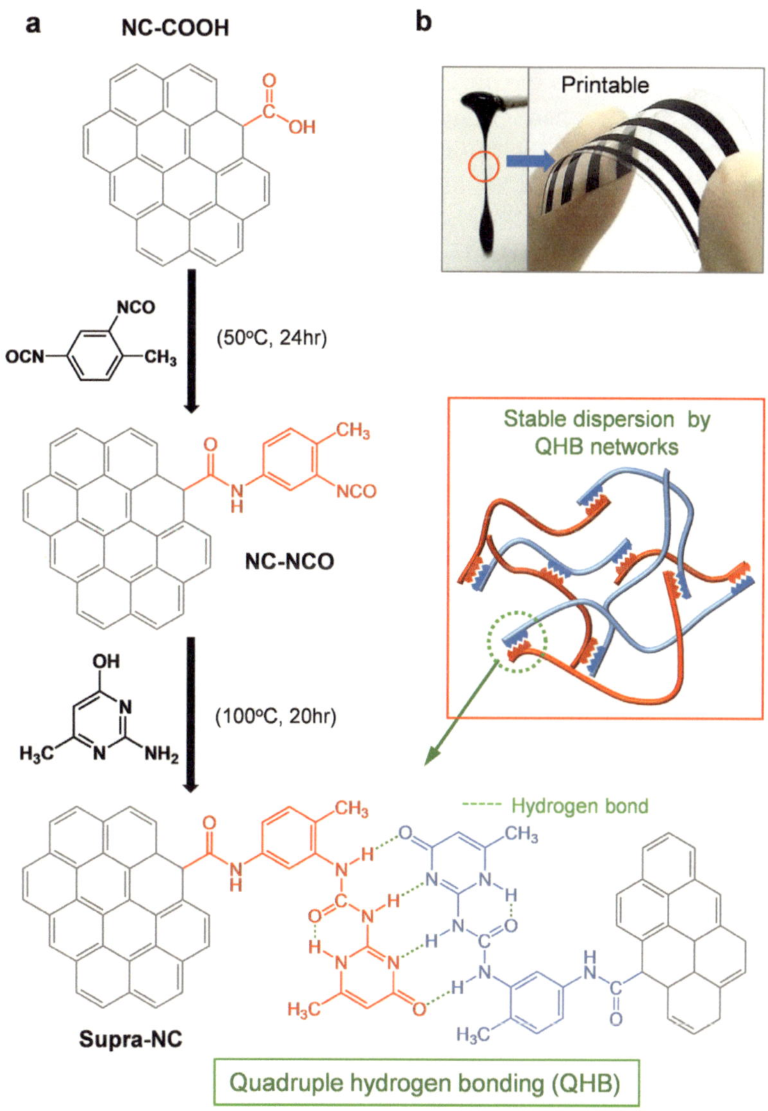

Figure 7. (a) Synthetic scheme for fabrication of graphene nanosheets functionalized with 2-ureido-4[1*H*]pyrimidinone moieties via a sequential reaction with toluene diisocyanate (TDI) and 2-amino-4-hydroxy-6-methyl-pyrimidine (AHMP) to form. (b) Photographs of well-dispersed RGO paste and printed electrode on the plastic substrate [19].

4. Graphene oxide nanosheets for surface and interfacial engineering

4.1. Atomically thin graphene oxide as a multifunctional dopant

Highly oxidized GO with electron-withdrawing groups can be utilized as a strong p-type dopant of nanocarbon materials such as carbon nanotubes (CNTs) and graphene because of the charge transfer interactions between sp^2 carbon and the oxidative functional groups in GO [20, 21]. Doping nanocarbon materials with GO nanosheets have advantages: stable and strong p-doping that maintains the intrinsic properties of pristine CNT films and chemical vapor deposition (CVD)-graphene. Controlling the surface wetting properties of nanocarbon films is very important for their use in optoelectronic devices, which are fabricated by layering a hydrophilic material on top of hydrophobic carbon electrodes. Moreover, deposited GO nanosheets on porous CNT networks can reduce the surface roughness of the film. Further, it is worth noting that the doping state assisted by GO nanosheets is stable for more than 40 days at room temperature and atmospheric pressure compared to that doped with nitric acid. **Figure 8** illustrates the advantage of GO nanosheets as p-type dopants for CVD-graphene. Graphene oxide doping decreased the sheet resistance of CVD-graphene from 600 to 292 Ω/sq. The doping effect of GO nanosheets on the CVD-graphene was demonstrated using Kelvin probe force microscopy (KPFM) and Raman spectroscopy results. The KPFM images associated with AFM images show that the surface potential of the graphene/single GO sheet is negatively shifted by 120 mV. The bright region in the Raman map of the 2D peak shows a p-doped area in a single GO nanosheet. The gate-dependent I-V characteristics of CVD-graphene and GO-doped CVD-graphene show that the hole mobility of CVD-graphene is almost unaffected by doping. The hole mobility of GO-coated graphene was found to be 3330 cm^2/Vs, which is only slightly lower than that of pristine graphene (under equivalent device positions before GO coating), 3500 cm^2/Vs. Graphene oxide nanosheets can be also used to modify the properties of single-walled carbon nanotube networks by p-doping, flattening the network surface, and making it hydrophilic. This is useful for fabricating optoelectronic devices onto GO modified graphene or single-walled carbon nanotube (SWCNT) films.

4.2. Graphene oxide as an interfacial modifier

For the fabrication of SWCNT patterns on hydrophilic substrates, partially reduced GO nanosheets are used as interfacial adhesive layers on hydrophilic SiO_2 surfaces. Hydrophobic materials can be easily detached from hydrophilic substrates. Thus, to obtain stable interfacial structure, hydrophilic substrates are usually treated with surface modifiers such as silane coupling agents. In this context, the deposition of GO onto substrates and its partial reduction has several advantages. The partially reduced GO having hydroxyl and carboxyl groups can play as a role of the interfacial adhesive between the substrate and the deposited materials. Moreover, this process is scalable and straightforward because uniform SWCNT networks can be formed even by spraying on plastic substrates. In terms of optoelectronic device application, partially reduced GO can be used for the work function engineering with the conducting and semiconducting materials. Uniform GO films and patterns can be fabricated by blow-assisted

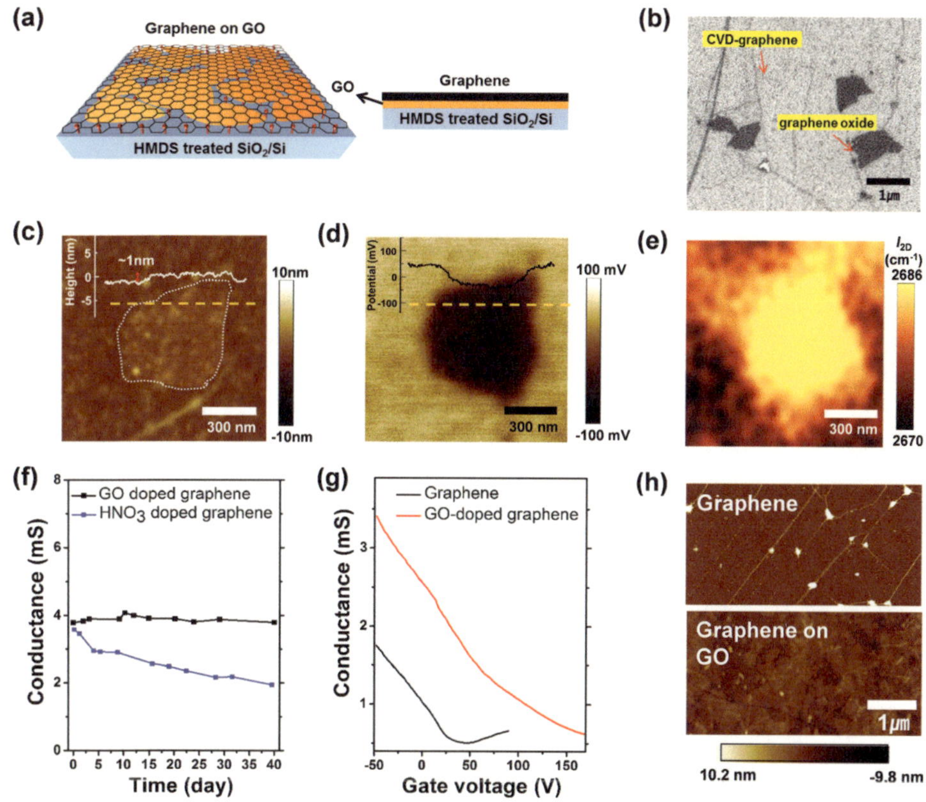

Figure 8. (a) Schematic diagram of graphene/GO film fabricated on an HMDS-treated SiO_2/Si substrate. (b) SEM image of CVD-grown graphene transferred on a GO sheet. (c) Atomic force microscope image and height profile showing the thickness of GO nanosheet, (d) kelvin probe force microscopy, and (e) Raman map of the 2D-band shift of CVD-grown graphene on a GO single sheet. (f) Electrical conductance variation of GO- and HNO_3-doped graphene with time at atmospheric pressure and room temperature. (g) Gate-dependent I-V characteristics of CVD-grown graphene and GO-doped graphene. (h) AFM images of CVD-grown graphene with and without GO [21].

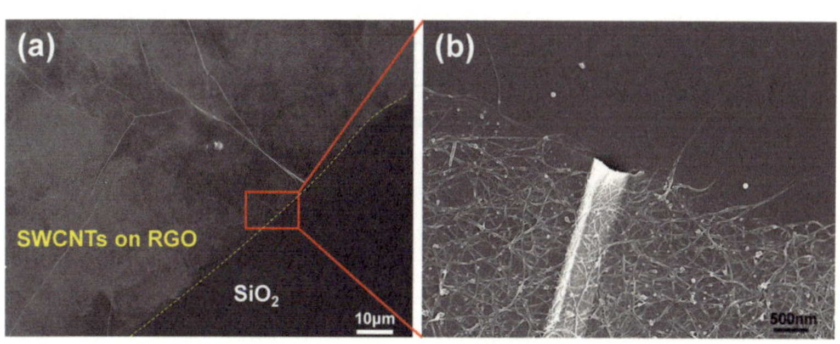

Figure 9. FESEM images of selectively deposited single-walled CNT films on partially reduced GO surfaces [22].

spin coating and inkjet printing, respectively, and the surface energy of the GO surface can be modulated by thermal treatment in vacuum. The SWCNTs were selectively deposited onto partially reduced GO films with moderately hydrophobic properties as shown in **Figure 9**.

5. Applications of graphene nanosheets in devices

5.1. Hierarchical graphene structure for lithium ion batteries

The hierarchical three-dimensional (3-D) structure for lithium ion battery (LIB) anodes has great potential for high electrochemical performances such as high-power densities and enhanced Coulomb's rates (C-rates). The efficient monolithic structure has a large specific surface area with numerous active sites, stable multistacking with short diffusion length, and high percolation threshold with high electrical conductivity as shown in **Figure 10(a)**. Unlike the porous-like graphene structuring on anode described in previous studies, monolithic graphene is similar to densely branched pine trees as shown in **Figure 10(b)**. The structure has high mechanical strength and flexibility, as well as high adhesion stability on the current collector. **Figure 10(c)** shows the growth mechanism for the structure by freeze-drying with water-soluble polymer [23]. The monolithic graphene anodes induced ultrafast charge/discharge rates with outstanding cycling stability with high capacitance as seen in **Figure 10(d)** and **(e)**. The fabrication method is simple and straightforward and suitable for high performance LIB anodes.

5.2. Fabrication of graphene oxide-based mid-IR detectors

Graphene oxide can be defined as chemically functionalized graphene containing oxygen. Moreover, GO has a large bandgap, which implies an insulating behavior. The bandgap can be decreased with decreasing the oxidation level of GO. Thus, this is a promising method for bandgap tuning that can be applicable to various optical and electronic devices. The spectrum ranges of GO-based photodetectors are limited within the visible and near-infrared (IR) wavelengths because they are based on the photovoltaic effect. Therefore, a new mechanism for mid-IR detection using GO sensors is required. Significantly, the control of oxygen functional

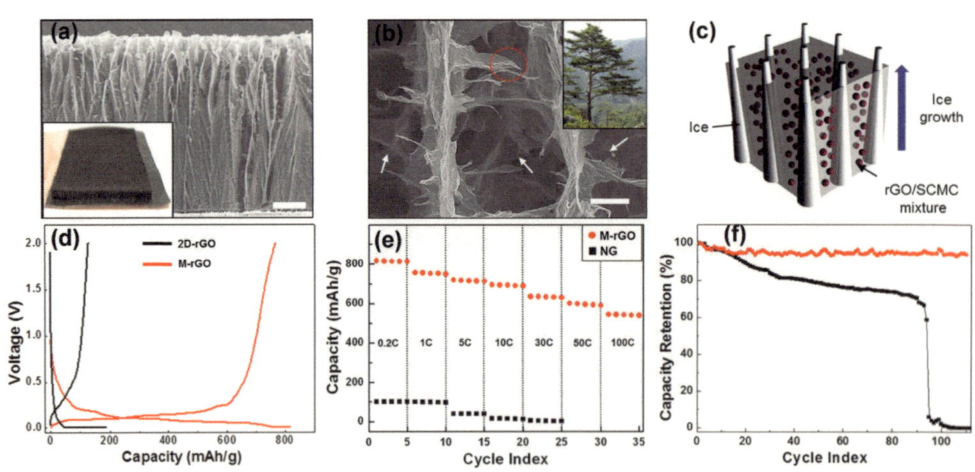

Figure 10. (a) FE-SEM image for monolithic RGO structure (inset: Vertically aligned RGO structure on current collector), (b) high magnification view of monolithic RGO, (c) growth mechanism for the monolithic structure during the freeze-drying process, (d) voltage profiles for monolithic-MGO (M-MGO) and 2-D RGO, (e) charge/discharge capacities for five cycles as the capacity rates increase from 0.2 to 100 C with respect to natural graphite and M-RGO, and (f) capacity retention for M-RGO (red line) and natural graphite (black line) [24].

groups in GO is possible for the application of IR detectors. (Bae et al. [25]) Here, GO is synthesized using thermal chemical vapor deposition (TCVD) and its electrical and optical properties are characterized using low-temperature measurements and Raman spectroscopy. The electrical conductivity of GO-based devices under IR irradiation was subsequently measured. The electrical characterization process can be described as follows. Single layered graphene (SLG) is synthesized on a copper substrate using the CVD method. Following its synthesis, the graphene is transferred onto SiO_2 substrate as shown in **Figure 11(a)**. Subsequently, SLG is deposited on a silicon substrate patterned using oxygen plasma etching. Furthermore, the metal electrode is deposited on the patterned substrate using Au as shown in **Figure 11(b)**. To produce GO, we performed chemical treatments using immersion in an aqueous acid solution as shown in **Figure 11(c)**. Subsequently, GO is cleaned with water to remove the residual acid and dried in vacuum as shown in **Figure 11(d)**. It should be noted that, in order to control the oxygen functional groups with certain resistance, it is reduced by annealing at optimum temperature as shown in **Figure 11(e)**. The optical microscopy image of the GO device is shown in **Figure 11(f)**. The dotted area indicates the fabricated GO channel between the metal electrodes.

Raman spectra illustrate the chemical doping effect by charge transfer between graphene and oxygen molecules. The spectra revealed significant changes in intensity ratio that can be described by the I(D)/I(G) and I(2D)/I(G) ratios. Moreover, the G band at 1590 cm^{-1} can be shifted due to the oxidation treatment. Four different samples: pristine CVD graphene, acid treated graphene, and annealed graphene after acid treatment are shown in **Figure 12(a)**. Initially, the I(D)/I(G) ratio increased after acid treatment compared to that of pristine graphene. This is caused by oxygen moieties generating intervalley scattering. After mild reduction by thermal treatment, the ratio decreases due to the partial elimination of oxygen functional groups. Thus, Raman spectra illustrate the doping effect of the oxygen functional groups. Moreover, charge transfer behavior presents as a p-type characteristic. The I(2D)/I(G) ratio is significantly decreased and the G band shows a red shift. Unlike graphene, GO has characteristic IR absorption peaks at

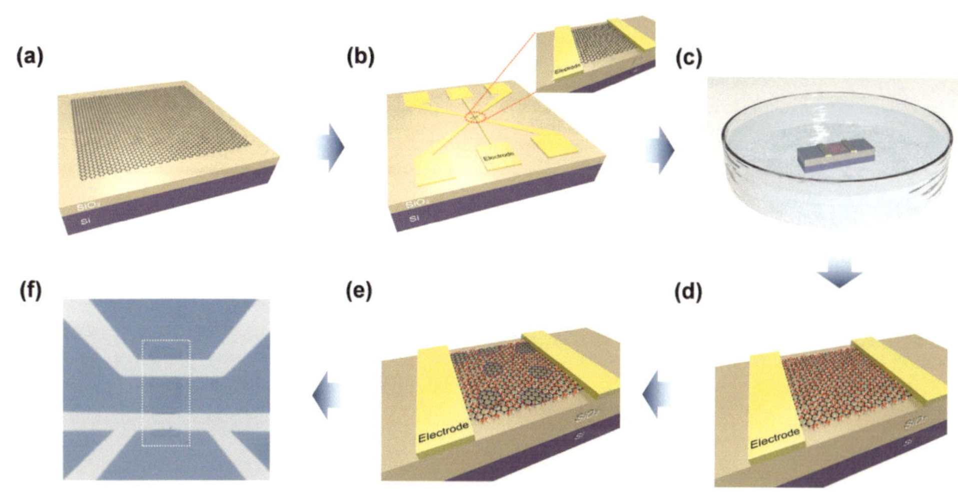

Figure 11. Fabrication schematics of the GO device: (a) transfer of CVD-grown graphene onto the Si/SiO_2 substrate, (b) patterning of graphene and deposition of electrodes, (c) acid treatment, (d) rinsing and drying, (e) reduction of GO, and (f) optical image of GO device [25].

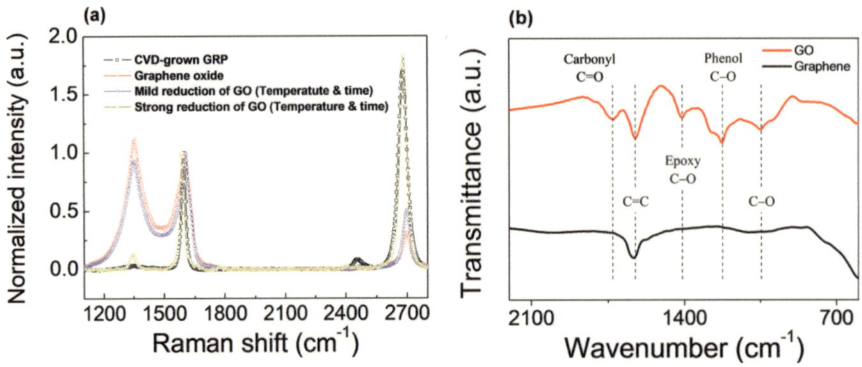

Figure 12. (a) Raman spectra of graphene, GO, and RGO and (b) IR absorption spectra of graphene and GO [25].

750–2250 cm^{-1}, as shown in **Figure 12(b)**. The Fourier transform infrared (FTIR) spectra of GO reveal oxygen functional groups from 960 to 1860 cm^{-1}, without graphene [26].

The optical properties of GO show in the mid-IR range at 7–14 μm. Depending on the functionality of GO by control of mild reduction, it reveals larger resistance changes as temperature. When the IR source is turned on, the resistance of the GO immediately drops from 31 to 7.4 MΩ, as shown in **Figure 13(a)**. This is a promising approach for obtaining sensitive mid-IR sensors by controlling GO functionalities. To describe the response of GO as a function of IR irradiation, the curve is fitted using nearest neighbor hopping (NNH) model as shown in **Figure 13(b)** and **(c)**. **Figure 13(c)** reveals a linear fit of ln(G)-T^{-1} in the 200–350 K temperature range, where G

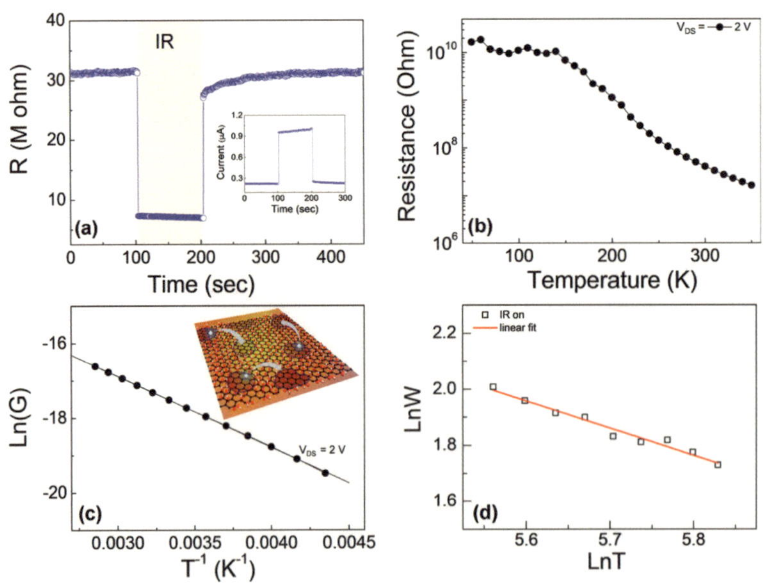

Figure 13. Electrical transport properties of GO under IR irradiation: (a) change in GO resistance, (b) GO resistance-temperature plot between 50 and 350 K, (c) temperature dependence of GO electrical conductance plotted for the NNH mechanism between 260 and 350 K, (d) ln(W) vs. ln(T) plot for GO [25].

http://dx.doi.org/10.5772/intechopen.77284

is the conductance and T is the temperature. Consequently, under IR irradiation, NNH is more dominant than variable range hoping (VRH) in the electron transport of GO, which is usually observed at a higher temperature range in a disordered system, jumping electrons to other defect sites due to thermal activation. Therefore, the different nature of the electron transport induced by IR irradiation contributes to the different temperature dependence exhibited by *G*. The activation energy *W*, extracted from the slope of the plot in **Figure 13(c)** is approximately 164 meV. The energy is easily obtained by the incident IR. The specific NNH transport phenomena is confirmed by plotting ln(*W*) as a function of ln(*T*) as shown in **Figure 13(d)**. The temperature exponent is −0.97, indicating thermal transport in GO under IR irradiation. The plot exhibits a negative slope between 250 and 350 K. These results demonstrate that the conducting mechanism can be attributed to VRH. The shift from VRH to NNH transport occurs due to the increase in the temperature of GO due to the incident IR irradiation.

5.3. Graphene emitters

Controlling the structure of graphene emitters such as the aspect ratio, density, and alignment, which are of practical importance for applications in field emission devices, is not readily achievable. Jeong et al. introduced a simple method for fabricating tubular structured graphene arrays with controlled tube lengths and alignment as shown in **Figure 14**. They used the filtration of RGO suspensions with a polycarbonate membrane. The interactions between hydrophobic graphene and the pore walls, but not the top surfaces, of a polycarbonate membrane were tuned, and the filtration rate was varied to control the length and alignment of the graphene arrays. They observed that the lengths of the graphene arrays increased with increasing filtration rate, but a maximum field emission efficiency was reached for intermediate filtration rates due to field screening for array lengths longer than an optimum value. The turn-on field and field enhancement factor for an optimum length of graphene arrays were 1.89 V/μm and 4624, respectively.

Another useful approach for fabricating graphene field emitters is a thermal welding-peeling method as shown in **Figure 15**. The graphene film was formed on a polytetrafluoroethylene membrane by filtering the dispersed RGO solutions. The CNT/polyethersulfone (PES)

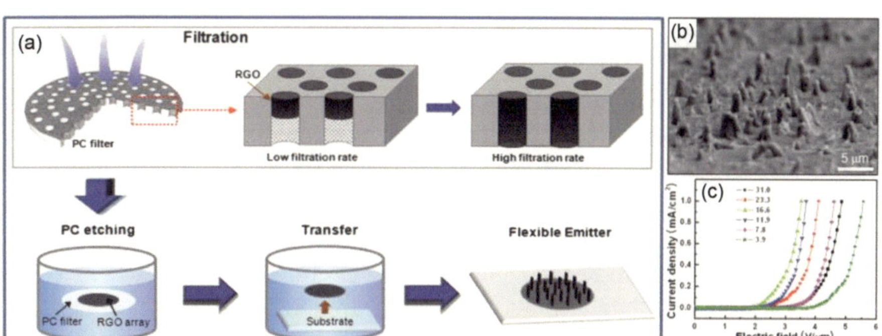

Figure 14. (a) Schematic and (b) SEM image of RGO emitters fabricated by filtration-transfer method. (c) J-E curves of RGO emitters as a function of filtration rate [27].

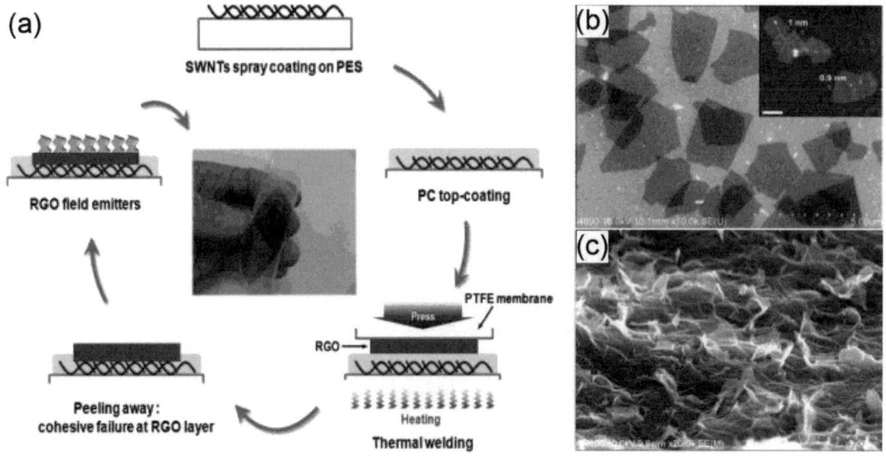

Figure 15. (a) Schematic diagram of RGO emitters fabricated using a thermal welding-peeling technique. SEM image of (b) RGO sheets and (c) fabricated RGO emitters [28].

substrate used as cathode was then coated with an adhesive polycarbonate layer to prepare an upper graphene thin film under pressed thermal treatment of the sample. Vertically aligned graphene emitters were finally constructed by peeling off the polytetrafluoroethylene membrane from the welded sample. The average height and interspace of graphene emitters were 1.2 and 0.8 μm, respectively, which could be controlled by experimental conditions such as the size of the GO sheets and peeling force of the membrane. A large field enhancement factor for the RGO emitters was achieved by optimum vertical alignment of GO nanosheets, resulting in a high emission current density and a low turn-on field.

The "breath figure" technique is a simple and versatile self-assembly method for fabricating porous nanomaterial patterns with high regularity. Although highly ordered 3-D graphene assemblies with high porosity have been fabricated, their use for field emitters was not readily achieved because the flat or smooth surface structures of the graphene assemblies did not emit electrons. Vertically aligned graphene ordered structures for an efficient field emitter was first fabricated using the "breath figure" method as shown in **Figure 16**. Octadecylamine (ODA)-functionalized GO solution dispersed in toluene was uniformly coated onto a substrate by spin coating. The ODA-functionalized GO was self-assembled under high humidity conditions and presented high periodicity due to the surface energy difference between the water droplet and toluene. Moreover, the ODA-functionalized GO nanosheets tended to encapsulate water droplets and precipitate at the water–solution interface, thereby preventing coalescence of the water droplets. The patterned ODA-functionalized GO array structure was obtained after the complete evaporation of the toluene and water droplets. The structure of GO array, the size, shape, and homogeneity, was dependent on the viscosity of the ODA-functionalized GO solution, large hexagonally structured GO patterns were fabricated with solutions having a low viscosity. The vertically aligned tip structures were formed at the interfaces between pores. However, with highly viscous solutions, small spherical patterns generated without tip formation. The graphene array prepared using a 2.0 g/L GO solution displayed the lowest turn-on field of 2.04 V/μm of all the arrays prepared using various GO concentrations.

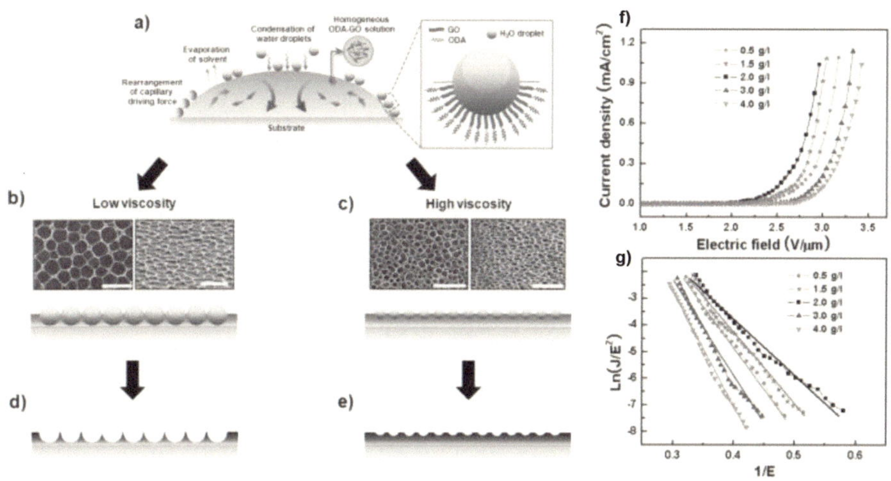

Figure 16. (a) Schematic diagram of the graphene arrays obtained using the 'breath figure" method, and pattern structures obtained at (b) and (d) low and (c) and (e) high solution viscosities, respectively. (f) J-E and (g) F-N plots of graphene arrays fabricated using various GO concentrations [29].

Freeze-drying of highly concentrated water-based RGO/polymer paste is one of the fabricating methods for 3-D graphene emitters with random micropores. Kim et al. first reported 3-D monolithic graphene structures as shown in **Figure 17**. They used highly concentrated water-based RGO paste prepared using monovalent cation-π interactions. After bar coating the paste on substrate, freezing was performed by immersing the sample in a liquid nitrogen bath. Low temperature freezing using liquid nitrogen resulted in the rapid formation of ice nuclei, hence the growth of relatively small ice crystals. After the sublimation of the ice crystals by vacuum drying, monolithic 3-D graphene structures with cylindrical pores could be obtained. Although the randomly distributed pores of the 3-D graphene structure were

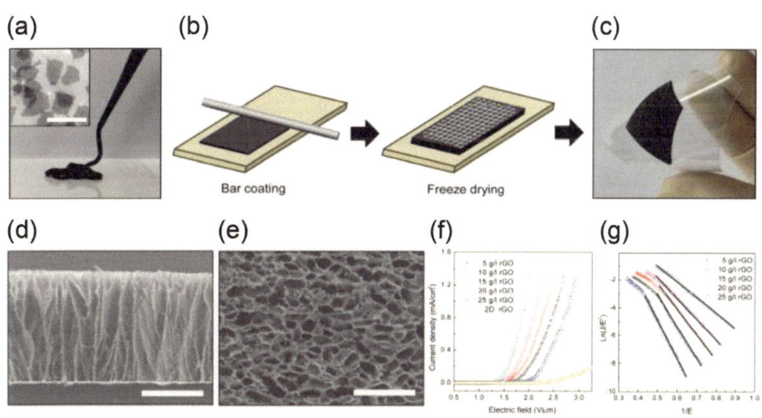

Figure 17. (a) Photo image of a water-based RGO paste. Inset shows the SEM image of the RGO sheets. The scale bar is 5 μm. (b) Fabrication method of the monolithic 3-D graphene structure using bar coating of the RGO paste, followed by subsequent freeze-drying. (c) a 3-D structured graphene fabricated on a plastic substrate. (d) Cross-sectional and (e) top view SEM images of a 3-D graphene structure. The scale bars are 300 μm. (f) J–E and (g) F-N plots of the 3-D graphene emitters as a function of graphene concentrations [30].

several tens of micrometers in size, the pore walls had numerous sharp edges. The size, shape, and homogeneity of these pores could be adjusted by choosing different freezing temperatures, solution concentrations, and solvents.

The tunneling barrier at the interface between a material surface and vacuum can be modulated by varying the physical properties of an emitter material. In this context, the field-emission characteristics are critically dependent on the work function of an emitter material. Low work functions decrease the barrier height causing the enhancement of electron tunneling for a given applied electric field, which results in high field-emission characteristics. Chemical doping can be a useful approach for modulating the work function of graphene because the intrinsic Fermi level of graphene can be readily shifted, due to charge transfer between the dopant and graphene. Jeong et al. reported the work-function engineering of graphene field emitters using chemical doping. Gold chloride as a p-type dopant and aluminum chloride powder as an n-type dopant were dissolved in distilled water and mixed with a GO solution. After centrifugation, the solution was filtered and the doped graphene films were dried. The schematic diagram in **Figure 18** shows the charge transfer from graphene to the gold and aluminum ions and the corresponding band diagrams. The charge transfer upon chemical doping was confirmed using various techniques such as Raman, X-ray photoelectron spectroscopy (XPS), and ultraviolet photoelectron spectroscopy (UPS). Due to decreasing the work-function of graphene, the Al-doped graphene emitters showed lower turn-on field than those of undoped and Au-doped graphene emitters. A similar study was conducted by mixing dopant solution with graphene paste.

Long-term emission stability of electron emitters is needed for thigh quality field emission devices. The ion bombardment of residual gas species which are degassed from cathodes,

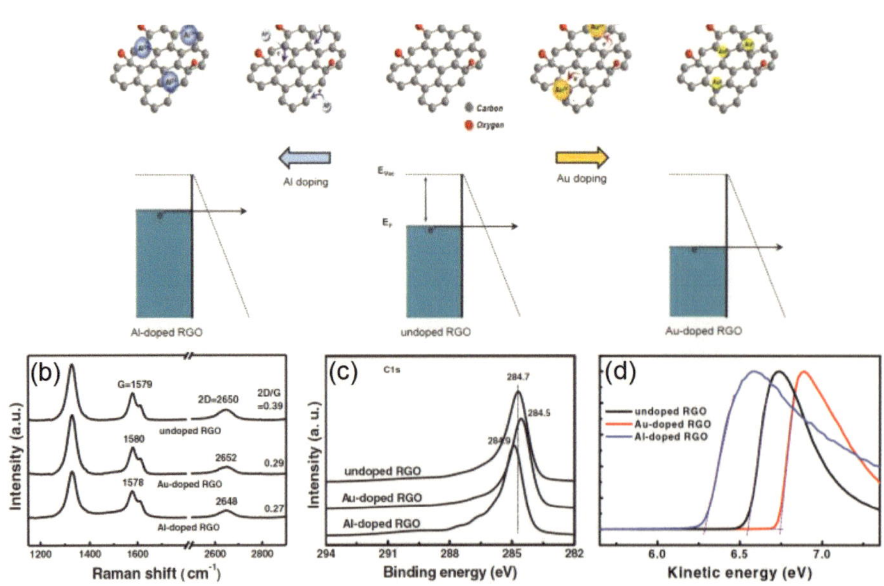

Figure 18. (a) Schematic diagram showing the electron transfer from graphene to the Au ion and from Al to graphene and corresponding band diagrams under application of a certain external voltage. (b) Raman, (c) XPS, and (d) UPS spectra for the undoped, Au-doped, and Al-doped graphene emitters [28].

getters, inner walls, and phosphors can destruct electron emitters, resulting in a critical reduction of the emitter characteristics. To enhance the current stability, Jeong et al. introduced the ZnO sol coating as a protective layer for the RGO emitters as shown in **Figure 19** [27]. Zinc oxide is an n-type semiconductor with a wide band gap and a low resistivity in the order of 10^{-2} to 10^{-3} Ω cm. The ZnO layer on the graphene surface was realized by hydrogen bonding between the amine groups of the ZnO sol and carboxyl groups of RGO and subsequent thermal treatment. A life time test showed stable emission for the ZnO-coated graphene emitters, which might be due to the ZnO protection of the emission site from reactive ion bombardment.

Since the development of graphene-based thin film fabrication techniques on polymeric substrates, research into graphene-based flexible electrodes for display application has advanced. In addition to the high electrical and mechanical flexibility of graphene-based thin films, the interface between an electrode and an emitter material should be strong and provide ohmic contact to achieve highly flexible field emitters. Thus, SWNT-coated polymer substrates have been used as electrodes [31]. Moreover, RGO emitters were fabricated on SWNT-coated PET substrates. A

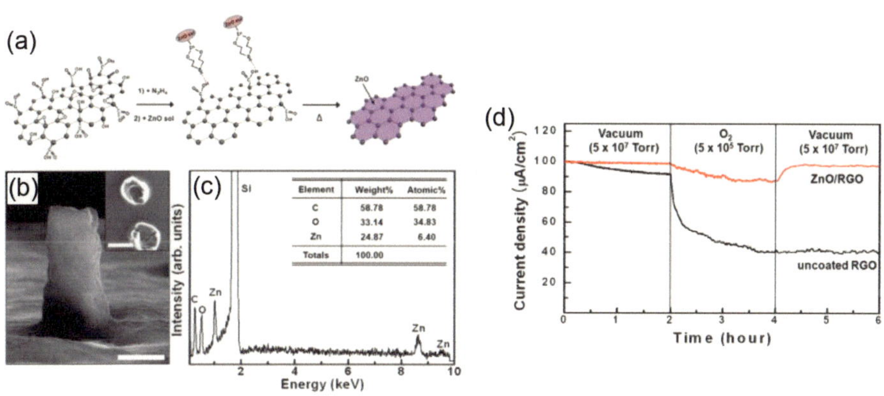

Figure 19. (a) Illustration of formation of a ZnO protective layer on the RGO surface. (b) SEM images and (c) EDAXS plot of the RGO emitters modified with a ZnO layer. Scale bars in (b) are 5 μm. (d) Current stability of the RGO array emitters wit and without a ZnO layer in vacuum and after exposure to O_2 [27].

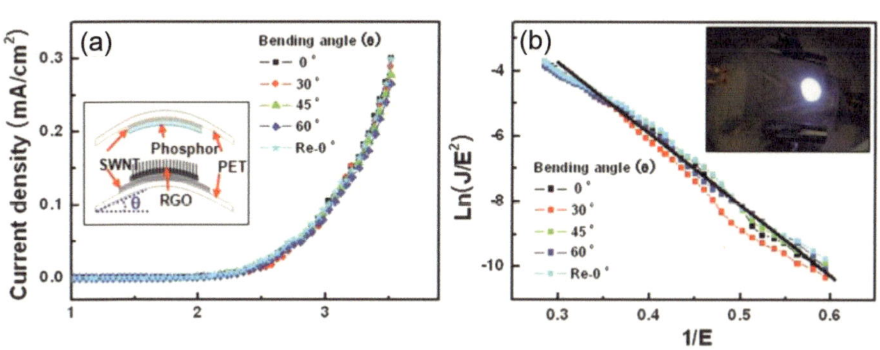

Figure 20. (a) J–E and (b) F–N plots of the RGO emitters as a function of the bending angle. The insets show schematic diagrams of the flexible field emission setup and an emission pattern at a 30° bending angle, respectively [26].

PET film was used as the spacer and a white phosphor-coated SWNT layer on a PET substrate was used as the anode as shown in **Figure 20**. Strong π-π carbon bonds allowed the RGO arrays to form strong mechanical contact with the SWNT network. The emission current density of the RGO emitters did not decrease much with increasing the bending angle. This stable emission is likely due to the strong adhesion of the 3-D RGO emitters to the SWNT-coated PET substrate.

6. Summary

We have briefly reviewed the recent research progress on chemically exfoliated graphene nanosheets via graphite oxide exfoliation and chemical reduction. Efficient graphite oxide exfoliation methods were developed by using homogenizers for shearing in solution and unusual horn sonication for stable acoustic cavitation. These methods show promise for fabricate improved GO nanosheets for high performance RGO nanosheets for conductor or electrochemical electrode applications. Highly oxidized GO nanosheets were utilized for p-type doping of CNTs and graphene films as well as for surface energy modifications. The modulation of the surface energy of GO can also allow us to deposit hydrophobic materials on hydrophilic surfaces. Strategies for the stable dispersion of RGO nanosheets in solution included sol–gel chemistry, cation-π interaction, supramolecular chemistry, and so on. Both GO and RGO nanosheets can be used as mid-IR detector, field emitters, 0r as electrodes in energy storage devices. Although some fascinating results have been achieved in previous publications, studying the fundamental and practical properties of GO or RGO should continue because their properties are critically dependent on the oxidation process of graphite, exfoliation method, reduction, and so on. Future applications of chemically exfoliated graphene in soft electronics, nanostructure control and hybridization with other materials are yet a challenge for high performance in real-life applications.

Acknowledgements

This work was supported by the Center for Advanced Soft-Electronics as Global Frontier Project (2014M3A6A5060953) and Nano-Material Technology Development Program (2016M3A7B4021151) funded by the Ministry of Science, ICT and Future Planning and by the Primary Research Program (18-12-N0101-16/18) of the Korea Electrotechnology Research Institute.

Author details

Joong Tark Han*, Seung Yol Jeong, Hee Jin Jeong and Geon-Woong Lee

*Address all correspondence to: jthan@keri.re.kr

Korea Electrotechnology Research Institute, Republic of Korea

References

[1] Eda G, Unalan HE, Rupesinghe N, Amaratunga GAJ, Chhowalla M. Field emission from graphene based composite thin films. Applied Physics Letters. 2008;**93**:233502. DOI: 10.1063/1.3028339

[2] Loh KP, Bao Q, Eda G, Chhowalla M. Graphene oxide as a chemically tunable platform for optical applications. Nature Chemistry. 2010;**2**:1015-1024. DOI: 10.1038/nchem.907

[3] El-Kady MF, Strong V, Dubin S, Kaner RB. Laser scribing of high-performance and flexible graphene-based electrochemical capacitors. Science. 2012;**16**:1326-1330. DOI: 10.1126/science.1216744

[4] Dreyer DR, Park S, Bielawski CW, Ruoff RS. The chemistry of graphene oxide. Chemical Society Reviews. 2010;**39**:228-240. DOI: 10.1039/B917103G

[5] Brodie BC. On the atomic weight of graphite. Philosophical Transactions of the Royal Society of London. 1859;**149**:249-259. DOI: 10.1098/rstl.1859.0013

[6] Staudenmaier L. Verfahren zur Darstellung der Graphitsaure. Berichte der Deutschen Chemischen Gesellschaft. 1898;**31**:1481-1499. DOI: 10.1002/cber.18980310237

[7] Hummers JWS, Offeman RE. Preparation of graphitic oxide. Journal of the American Chemical Society. 1958;**80**:1339. DOI: 10.1021/ja01539a017

[8] Park S, Ruoff RS. Chemical methods for the production of graphenes. Nature Nanotechnology. 2009;**4**:217-224. DOI: 10.1038/nnano.2010.69

[9] Jeong SY, Kim SH, Han JT, Jeong HJ, Yang S, Lee G-W. High performance transparent conductive films with rheologically-derived reduced graphene oxide. ACS Nano. 2011; **5**:870-878. DOI: 10.1021/nn102017f

[10] Li JL, Kudin KN, McAllister MJ, Prud'homme RK, Aksay IA, Car R. Oxygen-driven unzipping of graphitic materials. Physical Review Letters. 2006;**96**:176101. DOI: 10.1103/PhysRevLett.96.176101

[11] Ajayan PM, Yakobson BI. Materials science: Oxygen breaks into carbon world. Nature. 2006;**441**:818-819. DOI: 10.1038/441818a

[12] Xu H, Zeiger BW, Suslick KS. Sonochemical synthesis of nanomaterials. Chemical Society Reviews. 2013;**42**:2555-2567. DOI: 10.1039/C2CS35282F

[13] Han JT, Jeong JI, Kim HK, Hwang JY, Yoo HK, Woo JS, Choi S, Kim HY, Kim HJ, Jeong HJ, Jeong SY, Baeg K-J, Cho K, Lee G-W. Extremely efficient liquid exfoliation and dispersion of layered materials by unusual acoustic cavitation. Scientific Reports. 2014;**4**:5133. DOI: 10.1038/srep05133

[14] Manga KK, Wang S, Jaiswal M, Bao Q, Loh KP. High-gain graphene-titanium oxide photoconductor made from inkjet printable ionic solution. Advanced Materials. 2010; **22**:5265-5270. ISSN 0935-9648

[15] Han JT, Kim BG, Yang M, Kim JS, Jeong HJ, Jeong SY, Hong SH, Lee GW. Titania-assisted dispersion of carboxylated single-walled carbon nanotubes in a ZnO sol for transparent conducting hybrid films. ACS Applied Materials & Interfaces. 2011a;**3**:2671-2676 ISSN 1944-8244

[16] Han JT, Kim BJ, Kim BG, Kim JS, Jeong BH, Jeong SY, Jeong HJ, Cho JH, Lee G–W. Enhanced electrical properties of reduced graphene oxide multilayer films by in-situ insertion of a TiO_2 layer. ACS Nano. 2011;**5**:8884-8891. DOI: 10.1021/nn203054t

[17] Jeong SY, Kim SH, Han JT Jeong HJ, Jeong SY, Lee G-W. Highly concentrated and conductive reduced graphene oxide nanosheets by cation-π interaction: Toward printed electronics. Advanced Functional Materials. 2012;**22**:3307-3314. DOI: 10.1002/adfm.201200242

[18] Han JT, Jang JI, Jeong BH, Kim BJ, Jeong SY, Jeong HJ, Cho JH, Lee G–W. Spontaneous reduction and dispersion of graphene nano-platelets with in situ synthesized hydrazine assisted by hexamethyldisilazane. Journal of Materials Chemistry. 2012;**22**:20477-20481. DOI: 10.1039/C2JM34691E

[19] Han JT, Jeong BH, Seo SH, Roh KC, Kim S, Choi S, Woo JS, Kim HY, Jang JI, Shin D-C, Jeong S, Jeong HJ, Jeong SY, Lee G-W. Dispersant-free conducting pastes for flexible and printed nanocarbon electrodes. Nature Communications. 2013;**4**:2491. DOI: 10.1038/ncomms3491

[20] Han JT, Kim JS, Jo SB, Kim SH, Kim JS, Kang B, Jeong HJ, Jeong SY, Cho K, Lee G-W. Graphene oxide as a multi-functional p-dopant of transparent single-walled carbon nanotube films for optoelectronic devices. Nanoscale. 2012;**4**:7735-7742. DOI: 10.1039/C2NR31923C

[21] Kim H, Kim HH, Jang JI, Lee SK, Lee G–W, Han JT, Cho K. Doping graphene with an atomically thin two dimensional molecular layer. Advanced Materials. 2014;**26**:8141-8146. DOI: 10.1002/adma.201403196

[22] Han JT, Kim JS, Kwak D, Kim BG, Jeong BH, Jeong SY, Jeong HJ, Cho K, Lee G-W. Transparent carbon nanotube patterns templated by inkjet-printed graphene oxide nanosheets. RSC Advances. 2011;**1**:44-47. DOI: 10.1039/C1RA00213A

[23] Zhang H, Hussain I, Brust M, Butler M, Rannard SP, Cooper AI. Aligned two- and three-dimensional structures by directional freezing of polymers and nanoparticles. Nature Materials. 2005;**4**:787-793. DOI: 10.1038/nmat1487

[24] Jeong SY, Yang S, Jeong S, Kim IJ, Jeong HJ, Han JT, Baeg KJ, Lee GW. Monolithic graphene trees as anode material for Lithium ion batteries with high C-rates. Small. 2015;**11**:2774-2781. DOI: 10.1002/smll.201403085

[25] Bae JJ, Yoon JH, Jeong S, Boon BH, Han JT, Jeong HJ, Lee G-W, Hwang HR, Lee YH, Jeong SY, Lim SC. Sensitive photo-thermal response of graphene oxide for mid-infrared detection. Nanoscale. 2015;**7**:15695-15700. DOI: 10.1039/C5NR04039F

[26] Acik M, Lee G, Mattevi C, Chhowalla M, Cho K, Chabal YJ. Unusual infrared-absorption mechanism in thermally reduced graphene oxide. Nature Materials. 2010;**9**:840-845. DOI: 10.1038/nmat2858

[27] Jeong HJ, Jeong HD, Kim HY, Kim SH, Kim JS, Jeong SY, Han JT, Lee G-W. Flexible field emission from thermally welded chemically doped graphene thin films. Small. 2012;**8**:272-280. DOI: 10.1002/smll.201101696

[28] Jeong HJ, Kim HY, Jeong HD, Jeong SY, Han JT, Lee G-W. Arrays of vertically aligned tubular-structured graphene for flexible field emitters. Journal of Materials Chemistry. 2012;**22**:11277-11283. DOI: 10.1039/C2JM31263H

[29] Jeong HJ, Jeong HD, Kim HY, Jeong SY, Han JT, Lee G-W. Self-organized graphene nanosheets with corrugated, ordered tip structures for high-performance flexible field emission. Small. 2013;**9**:2182-2188. DOI: 10.1002/smll.201202143

[30] Kim HY, Jeong S, Jeong SY, Baeg K-J, Han JT, Jeong MS, Lee G-W, Jeong HJ. Chemically doped three-dimensional porous graphene monoliths for high-performance flexible field emitters. Nanoscale. 2015;**7**:5495-5502. DOI: 10.1039/C4NR07189A

[31] Jeong HJ, Jeong HD, Kim HY, Kim JS, Jeong SY, Han JT, Bang DS, Lee G-W. All-carbon nanotube-based flexible field-emission devices: From cathode to anode. Advanced Functional Materials. 2011;**21**:1526-1532. DOI: 10.1002/adfm.201001469